Tenth Edition

Basic Laboratory Studies
in GENERAL CHEMISTRY
with Semimicro Qualitative Analysis

Grace R. Hered
City Colleges of Chicago

Houghton Mifflin Company
Boston   New York

Senior Sponsoring Editor: Richard Stratton
Assistant Editor: Marianne Stepanian
Director of Manufacturing: Michael O'Dea
Executive Marketing Manager: Karen Natale

Cover Design: Harold Burch, Harold Burch Design, New York City.
Cover Image: Peter Arnold, Inc., © Manfred Kage, Vitamin B-12 crystal.

Printed in the U.S.A.

ISBN: 0-669-35491-0

2 3 4 5 6 7 8 9 - KP - 2008

Atomic Masses of the Elements (based on carbon-12 with the uncertainties in parentheses)

Name	Symbol	Atomic Number	Atomic Mass	Name	Symbol	Atomic Number	Atomic Mass
Actinium*	Ac	89	227.028	Molybdenum	Mo	42	95.94(1)
Aluminum	Al	13	26.981539(5)	Neilsbohrium*	Ns	107	(262)
Americium*	Am	95	(243)ᵃ	Neodymium	Nd^g	60	144.24(3)
Antimony (stibium)	Sb	51	121.75(3)	Neon	Ne^{g,m}	10	20.1797(6)
Argon	Ar^{g,r}	18	39.948(1)	Neptunium*	Np	93	237.048
Arsenic	As	33	74.92159(2)	Nickel	Ni	28	58.6934(2)
Astatine*	At	85	(210)ᵃ	Niobium	Nb	41	92.90638(2)
Barium	Ba	56	137.327(7)	Nitrogen	N^{g,r}	7	14.00674(7)
Berkelium*	Bk	97	(247)ᵃ	Nobelium*	No	102	(259)ᵃ
Beryllium	Be	4	9.012182(3)	Osmium	Os^g	76	190.23(3)
Bismuth	Bi	83	208.98037(3)	Oxygen	O^{g,r}	8	15.9994(3)
Boron	B^{g,m,r}	5	10.811(5)	Palladium	Pd^g	46	106.42(1)
Bromine	Br	35	79.904(1)	Phosphorus	P	15	30.973762(4)
Cadmium	Cd^g	48	112.411(8)	Platinum	Pt	78	195.08(3)
Calcium	Ca^g	20	40.078(4)	Plutonium*	Pu	94	(244)ᵃ
Californium*	Cf	98	(251)ᵃ	Polonium*	Po	84	(209)ᵃ
Carbon	C^r	6	12.011(1)	Potassium (kalium)	K	19	39.0983(1)
Cerium	Ce^g	58	140.115(4)	Praseodymium	Pr	59	140.90765(3)
Cesium	Cs	55	132.90543(5)	Promethium*	Pm	61	(145)ᵃ
Chlorine	Cl	17	35.4527(9)	Protactinium*	Pa	91	231.03588(2)
Chromium	Cr	24	51.9961(6)	Radium*	Ra	88	226.025
Cobalt	Co	27	58.93320(1)	Radon*	Rn	86	(222)ᵃ
Copper	Cu^r	29	63.546(3)	Rhenium	Re	75	186.207(1)
Curium*	Cm	96	(247)ᵃ	Rhodium	Rh	45	102.90550(3)
Dysprosium	Dy^g	66	162.50(3)	Rubidium	Rb^g	37	85.4678(3)
Einsteinium*	Es	99	(252)ᵃ	Ruthenium	Ru^g	44	101.07(2)
Erbium	Er^g	68	167.26(3)	Rutherfordium*	Rf	104	(261)
Europium	Eu^g	63	151.965(9)	Samarium	Sm^g	62	150.36(3)
Fermium*	Fm	100	(257)ᵃ	Seaborgium*	Sg	106	(263)
Fluorine	F	9	18.9984032(9)	Scandium	Sc	21	44.955910(9)
Francium*	Fr	87	(223)ᵃ	Selenium	Se	34	78.96(3)
Gadolinium	Gd^g	64	157.25(3)	Silicon	Si^r	14	28.0855(3)
Gallium	Ga	31	69.723(1)	Silver	Ag^g	47	107.8682(2)
Germanium	Ge	32	72.61(2)	Sodium (natrium)	Na	11	22.989768(6)
Gold	Au	79	196.96654(3)	Strontium	Sr^{g,r}	38	87.62(1)
Hafnium	Hf	72	178.49(2)	Sulfur	S^r	16	32.066(6)
Hahnium*	Ha	105	(262)	Tantalum	Ta	73	180.9479(1)
Hassium*	Hs	108	(265)	Technetium*	Tc	43	(98)ᵃ
Helium	He^{g,r}	2	4.002602(2)	Tellurium	Te^g	52	127.60(3)
Holmium	Ho	67	164.93032(3)	Terbium	Tb	65	158.92534(3)
Hydrogen	H^{g,m,r}	1	1.00794(7)	Thallium	Tl	81	204.3833(2)
Indium	In	49	114.818(3)	Thorium*	Th^{g,z}	90	232.0381(1)
Iodine	I	53	126.90447(3)	Thulium	Tm	69	168.93421(3)
Iridium	Ir	77	192.217(3)	Tin	Sn^g	50	118.710(7)
Iron	Fe	26	55.845(2)	Titanium	Ti	22	47.88(3)
Krypton	Kr^{g,m}	36	83.80(1)	Tungsten (wolfram)	W	74	183.85(3)
Lanthanum	La^g	57	138.9055(2)	Ununbium*	Uub	112	(277)
Lawrencium*	Lr	103	(260)ᵃ	Ununnilium*	Uun	110	(269)
Lead	Pb^{g,r}	82	207.2(1)	Unununium*	Uuu	111	(272)
Lithium	Li^{g,m,r}	3	6.941(2)	Uranium*	U^{g,m,z}	92	238.0289(1)
Lutetium	Lu^g	71	174.967(1)	Vanadium	V	23	50.9415(1)
Magnesium	Mg	12	24.3050(6)	Xenon	Xe^{g,m}	54	131.29(2)
Manganese	Mn	25	54.93805(1)	Ytterbium	Yb^g	70	173.04(3)
Meitnerium*	Mt	109	(266)	Yttrium	Y	39	88.90585(2)
Mendelevium*	Md	101	(258)ᵃ	Zinc	Zn	30	65.39(2)
Mercury	Hg	80	200.59(3)	Zirconium	Zr^g	40	91.224(2)

ᵃAtomic mass values in parentheses are used for radioactive elements the atomic masses of which are not known precisely or that cannot be quoted precisely without knowledge of the origin of the elements; the value given is the atomic mass number of the isotope of that element that has the longest known half-life.

^{g}Geological specimens are known in which the element has an isotopic composition outside the limits for normal material. The difference between the atomic mass of the element in such specimens and that given in this table may exceed the implied uncertainty.

^{m}Modified isotopic compositions may be found in commercially available material because it has been subjected to an undisclosed or inadvertent isotopic separation. Substantial deviations in atomic mass of the element from that given in the table can occur.

^{r}Range in isotopic composition of normal terrestrial material prevents a more precise mass being given; the tabulated mass value should be applicable to any normal material.

^{z}An element, without stable nuclide(s), exhibiting a range of characteristic terrestrial compositions of long-lived radionuclide(s) such that a meaningful atomic mass can be given.

*Element has no stable nuclides.

Adapted from *Pure and Applied Chemistry*, Vol. 66, pages 2426–2427 (1994).

dan
doug

Contents

Preface   vii

Contents

PART 3   Some Laboratory Skills and Devices   385

Preface

Basic Laboratory Studies in General Chemistry with Semimicro Qualitative Analysis,
Tenth Edition, follows closely in style and sequence the textbook material in *General
Chemistry* and *General Chemistry with Qualitative Analysis,* Tenth Editions, by
William R. Robinson. It is, however, independent enough that it may also be used with
other, comparable textbooks. A sufficient amount and variety of material is provided
for a two-semester course in general chemistry and qualitative analysis, but this
manual is also versatile enough for use in a oné-semester introductory course in
chemistry, a one- or two-semester course in chemistry for the health sciences, or a
one-semester course in semimicro qualitative analysis. Enough pertinent exercises are
included for a science course for nonscience majors or for environmentally oriented
majors.

The principles that have guided the nine previous editions have been adhered to
in this edition. These guiding principles are

1. the use of descriptive chemistry to illustrate the basic concepts of chemistry;
2. the nurturing of independent thought;
3. the employment of problem solving, quantitative operations, and identification of
   unknowns as prominent methods of teaching;
4. the choice of apparatus and materials according to their greatest availability,
   economy, and safety.

The format of the tenth edition is similar to that of the ninth edition. It has five major
parts: Part 1, Introduction; Part 2, Exercises; Part 3, Some Laboratory Skills and
Devices; Part 4, Semimicro Qualitative Analysis; and the Appendix.

It is essential that students read Part 1·before actually carrying out any laboratory
work. Part 1A contains general introductory information about the use of the manual
and working in the laboratory. The contents of Part 1B, "Safety in the Laboratory,"
must be studied and understood completely before the student begins the exercises. A
Laboratory Safety Quiz at the end of Part 1B should be completed to the satisfaction
of the instructor. (Students should also view any available video material about
laboratory safety.) The manual uses boldface type to emphasize the importance of
safety (see "Description of Hazardous Chemicals" in the Appendix). Part 1C,
"Preparing for the Laboratory," lists directions needed for the student to prepare in
advance for the actual laboratory time. Part 1D explains how to record data obtained
and how to use the Report forms.

Part 2 contains 45 laboratory-tested exercises. Each exercise is followed by
"Thought Questions," to be answered *outside* of laboratory time, and assigned prior
to the experiment. This practice will enhance student understanding of the exercise.
The order of the exercises follows in part the organization of material in the tenth
edition text by Robinson. The exercises, however, can be used in any order to fit an
instructor-designed presentation. Exercises 1 through 29 are essentially traditional

general chemistry experiments. Exercises 30, 31, and 32 stress basic principles applicable to environmental studies. Properties of Oxygen and Hydrogen (Exercise 33) has been maintained for nonscience majors and/or students who have had no previous chemistry experience. With such students, attention should be drawn to the Appendix, which includes directions for the laboratory preparation of common gases, if the need for these arises—namely, oxygen and hydrogen (also chlorine and bromine). A sequence of exercises applicable but not restricted to organic, biological, or health-oriented chemistry includes Exercises 34 through 42. Exercises 43, 44, and 45 deal with special topics.

Part 3, "Some Laboratory Skills and Devices," presents detailed information on essential laboratory techniques required to perform the experiments. Students should be aware of the availability of this information when certain techniques are needed for the first time. They should know they will need to refer to it from time to time.

"Semimicro Qualitative Analysis," Part 4, is adapted from *General Chemistry with Qualitative Analysis,* Tenth Edition. It presents the procedures that A. A. Noyes developed. Flow sheets for cation and anion group separations and some pertinent illustrations are included.

The Appendix contains reference data as well as explanatory information for the student in the following areas: Units and Conversion Factors, Dimensional Analysis, Accuracy and Precision, Significant Figures, Rounding Off Numbers, and Standard Exponential (Scientific) Notation. The Appendix also includes directions for the generation of common gases, a description of hazardous chemicals, and a section on the disposal of wastes. The list of materials needed for the exercises helps in planning and carrying out the work to be done each semester.

An instructor's guide for this manual is available. The *Instructor's Guide for Basic Laboratory Studies in General Chemistry with Semimicro Qualitative Analysis,* Tenth Edition, includes helpful suggestions for using the laboratory manual effectively, scheduling the laboratory program, and alerting students to the safety precautions. Resources for Part 2 of the laboratory manual include a list of suitable unknowns for the exercises, special notes for each exercise, answers to the Thought Questions, and a suggested list of locker equipment. For Part 4, the Instructor's Guide includes notes on the qualitative analyses, answers to the questions, and a suggested list of locker supplies. It also provides suggestions for grading reports, a comprehensive list of required chemicals and supplies, and a log for recording equipment and supplies available in the storeroom or laboratory.

This edition includes a complete review and some revision of procedural directions, and the maintenance of a variety of traditional experiments along with ones covering a range of selective topics.

Grateful acknowledgment for professional advice is made to the numerous teachers, critics, and students who have assisted in this project.

Grace R. Hered

Part 1

Introduction

A

To the Student

Chemistry, the study of matter, is a laboratory science. Experimentation, carried out in the chemical laboratory, is the foundation of chemical knowledge. Before *any* laboratory work can be carried out, safety in the laboratory must be thoroughly understood, so that *all* accidents are prevented. Carefully read the following information on this subject contained in the *Laboratory Regulations* and *Safety Rules and Procedures* sections. Then take the *Laboratory Safety Quiz* (pages 9 and 10),

which is on a removable report sheet. Turn in the completed quiz to your instructor, if he or she so directs. If this quiz has been completed satisfactorily, you are on your way to beginning actual laboratory work. Your next obligation, which can be accomplished outside the laboratory itself, is to read and study Parts 1C and 1D, *Preparing for the Laboratory* and *Report Writing,* respectively (pages 11–15).

B

Safety in the Laboratory

1. Laboratory Regulations

The following regulations must be observed for efficiency and safety in the laboratory. Your own laboratory may have special regulations of which you will be made aware.

1. **Perform only those exercises that are authorized.**
2. **Follow the directions in the assigned exercise exactly. Carefully check the identity of chemicals used in the exercise, and their specified concentrations.**
3. **Never work alone in an unsupervised laboratory.**
4. Never throw solid materials into sinks; use the proper waste containers for paper and glass. Some chemical solid and liquid wastes will have to be disposed of in special containers, as indicated by the instructor.
5. Do not insert spatulas or pipets into reagent bottles. Instead, transfer an approximate amount of the reagent desired to a glazed paper, a watch glass, or a test tube. Do not carry stock bottles to your desk. Do not return excess to stock containers, as this may contaminate the stock material.
6. Weigh solids on glazed paper or on a watch glass; do not allow chemicals to come into contact with balance pans.
7. Clean up spilled materials immediately, using liberal quantities of water. Ask the instructor for assistance if strong acids or strong bases have been spilled. Refer to *Disposal of Wastes* in the Appendix, where necessary.
8. Keep your working surfaces clean at all times. Use a damp cloth or sponge for wiping the surface and a towel for drying.
9. Keep floor area dry and clear of glass or other hazards that could cause slipping. ·

10. Before leaving the laboratory, make sure the water and gas are completely shut off. Return all special equipment to the stockroom or designated place.

2. Safety Rules and Procedures*

Familiarize yourself with the *location* of exits, fire extinguishers, showers, fire blankets, and eyewashes. If first aid charts are posted, study them at the beginning of the course. Know the location of circuit breakers, which can turn off the electricity.

A typical carbon dioxide fire extinguisher (Fig. B.2.1), or dry-powder type, is used by breaking the wire or plastic seal and squeezing the hand grip while directing the nozzle at the *base* of the

Figure B.2.1   A typical carbon dioxide fire extinguisher.

*View a videotape on laboratory safety, if one is available. For more details, see *Safety in Academic Chemistry Laboratories*, 5th Ed., American Chemical Society, 1155 16th St., N.W., Washington, D.C., 20036, 1990, or similar available publication.

Figure B.2.2   A safety shower with pull chain.

flame. Any use of an extinguisher must be reported to your instructor so that it can be checked, refilled, and resealed, or replaced, for the next use.

A safety shower (Fig. B.2.2) is equipped with a pull chain that activates the shower head. Showers can wash off dangerous chemicals, and also extinguish a person's burning clothes.

A fire blanket is designed to smother a clothing fire. One type is pulled out and wrapped around the body. Identify the type in your laboratory.

An eyewash facility (Fig. B.2.3) is essential in every laboratory, and some types are operated by pushing the valve to turn on the water. Determine which type your laboratory has, if any.

Report *all* accidents immediately to your instructor. Some accidents can destroy an eye or cause other serious consequences in a few seconds; other accidents can have delayed effects.

If you are aware of any potential or impending hazard, notify your instructor immediately.

A. Eye Protection. Wear safety goggles in the laboratory *at all times* when experiments are being carried out, as directed by your instructor. Safety goggles can be worn over prescription glasses.

Contact lenses are *not* permitted in the laboratory. Chemicals splashed into the eye may become trapped under hard lenses or absorbed by soft

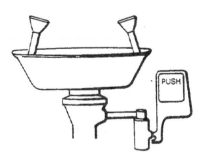

Figure B.2.3   An eyewash facility.

lenses. It is difficult to remove contact lenses before washing injured eyes with water. Newer lenses are permeable to organic fumes.

When you are heating substances in a test tube, never look down into it nor point its mouth toward anyone. It may erupt and spray its contents, perhaps into the eyes. Gradual heating over the surface of the *slanted* tube rather than in one spot and the addition of a small boiling chip to liquids minimize this danger. Never heat liquids in graduated cylinders—they will crack.

If a chemical gets in the eyes, first aid consists of thorough washing of the eyes with water from a laboratory eyewash facility for 10 to 15 minutes. Avoid a high-pressure stream of water, so as not to injure delicate eye tissue. If no eyewash facility is available, place the injured person on his or her back and, holding the eye(s) open, pour water gently into the eye(s)—again for 10 to 15 minutes. Prompt medical attention must be given, regardless of the severity of the injury.

B. Fire Precautions. Be on constant alert to minimize the possibility of fires. Keep the workspace clear of flammable materials such as excess paper or articles of clothing. Do not wear loose, flowing clothing. Tie back long hair. Students with beards, take care!

Never smoke in the laboratory.

Immediately extinguish small chemical fires such as may occur in beakers, crucibles, or test tubes by covering with a crucible or another beaker, if a ceramic-centered gauze is not available.

Do not heat flammable liquids with an open flame; use a hot plate or steam bath.

For larger fires, use the fire extinguisher, as described earlier. Gas and electricity must be turned off by the instructor, or any available person. Clothing fires are treated by using the fire blanket, the shower, dousing with water, or rolling the person on the floor—whichever is most expedient.

In case of an electrical fire, use the circuit breaker, and proceed to extinguish the fire.

First aid for minor heat burns consists of immersing or covering the affected area with ice water. **Do not apply salve or oils to a burn.** For serious burns, or any injury, see a physician immediately.

C. Chemical Burns, Poison, and Fume Hazards. For protection against chemical spills, wear suitable footwear (no open sandals) and an apron in the laboratory at all times.

In the event of contact with a corrosive chemical, wash it away with large amounts of cold running water. Remove any clothing that may have been contaminated. Once all the chemicals have been washed away, if a chemical burn has occurred, treat it as if it were a heat burn; namely, with ice-cold water (**no salve or oils**). For burns exhibiting any visual damage, see a physician immediately.

· Do not eat, drink, or prepare food in the laboratory. Do not store food in a refrigerator that is used to store chemicals. Do not use laboratory glassware, such as beakers, for drinking glasses. Never taste a chemcal.

In pipetting toxic chemicals, use an appropriate rubber safety bulb or pipet pump (Fig. B.2.4) rather than mouth suction. Avoid contamination of the reagent by not aspirating it into the bulb or pump.

Avoid possible poisoning by contamination through the hands. Always wash hands thoroughly after exposure to hazardous chemicals and, as a general rule, after each laboratory period.

Whenever toxic, irritating, or flammable gases are likely to be evolved, the experiment must be conducted in the fume hood (Fig. B.2.5), a standard piece of laboratory equipment that exhausts these gases to the outside when the fan is turned on.

Never directly inhale vapors. If an odor must be identified, gently waft some of the vapor toward you, but do not breathe deeply.

D. Cuts. Do not use chipped, cracked, or broken glass equipment.

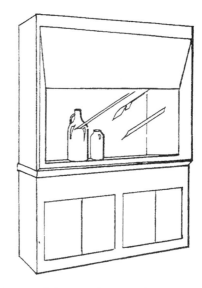

Figure B.2.5   A laboratory fume hood.

Exercise caution in using glass tubing. When inserting glass tubing into stoppers, or removing it from stoppers, use the following directions.

INSERTING GLASS TUBING IN A STOPPER

Make sure that both ends of the tubing have been fire-polished. Wet both the tubing and the hole in the stopper with one or two drops of water or glycerol. **Protecting your hand with a towel, and grasping the tubing near the stopper (not at far end), insert tubing into the hole in the stopper with a gentle twisting motion. Never try to** *push* **the tubing through the stopper, as the tubing is almost certain to snap and inflict serious cuts.** When glass thistle tubes are inserted into stoppers, the additional precaution must be taken *not* to hold the thistle tube by the enlarged portion while twisting.

REMOVING GLASS TUBING FROM A STOPPER

Wrap the tubing with a towel and pull it from the stopper with a gentle twisting motion. One or two drops of water or glycerol introduced around the hole and on the tubing makes removal easier. In stubborn cases, see the instructor.

E. Reagent Handling. Review copies of material safety data sheets on possibly hazardous chemical reagents that you may use, if your instructor directs.

Use only glassware that is initially clean. Contaminants may obscure the desired results of the experiment or may interact dangerously with the added material.

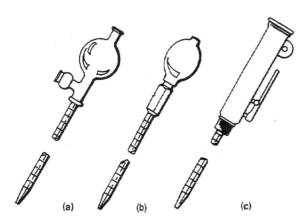

Figure B.2.4 Safety bulbs for pipets, (a) and (b). Pipet pumps (c) are also available.

Carefully read all labels on reagent bottles before using; an error may have serious consequences. **Use only the quantities and concentrations called for.**

Never change quantities or substitute chemicals unless specifically directed to do so by your instructor.

Never pour water into concentrated acid. Always pour the acid slowly into the water while stirring the mixture constantly. Cooling by a water bath may be necessary when diluting large quantities of acid or when diluting concentrated acids. Ventilation is important when working with acids and/or bases.

3. Laboratory Safety Quiz

NAME _Jason deYoung_ SECTION _____ DATE _8/26/09_

Safety in the chemical laboratory is extremely important. The results of this quiz will determine if you understand the material put forth under *Safety in the Laboratory*. Complete this quiz, to the satisfaction of your instructor, before you begin your laboratory work.

(1)

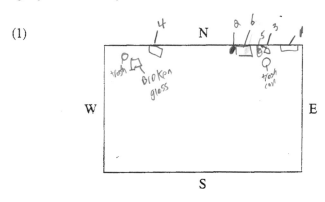

Show on the above diagram the location in your laboratory of the exits (*1*), fire extinguisher(s) (*2*), shower (*3*), fire blanket (*4*), eyewash facility (*5*), fume hood(s) (*6*), and waste receptacles (*7*). Indicate by placing the number in parenthesis following each of these terms in its proper location in the diagram. If an item is *outside* your laboratory, indicate by placing the number outside of the square, in its approximate location. One number may be used more than once.

(2) What is the most important part of the body to protect in the laboratory?

eye's

(3) How is the protection referred to in Question (2) achieved?

Wear safety goggles in the laboratory at all times when expaiments are being carried out.

(4) Why are contact lenses not permitted in the laboratory?

becaue chemical spilles con get trapped under the lenses, or absorbed

(5) How can small fires in beakers, etc., be put out?

by covering with crucible or another beaker

(6) After a fire extinguisher has been used, what must then be done by laboratory maintenance to the extinguisher, and why?

IT must be check, refilled, or replaced so it is ready for the next lab.

(7) How is a heat burn treated? How is a chemical burn treated?

both are treated the same, run cold water

(8) Why are there showers in laboratories?

So someone can wash chemical spills off there body

(9) Why must closed footwear and protective clothing be used in the laboratory?

because if there is a accident there is nothing protecting your feet.

(10) Name three ways chemical poisons can enter the body.

orillys by breathing it in, or obsorbing through the skin.

(11) What is a laboratory hood, and what is its use?

The hood is a protective area and its used for dealing with combostuble or flamable materials.

(12) What is the most common cause of cuts sustained in the laboratory?

broken or chipped glass from beakers and test tubes

(13) What substances can be used to assist in inserting glass tubing into stoppers?

water or glycerol

(14) Why is it important to carefully read all reagent-bottle labels?

So you know what your dealing with.

(15) Why should you not work alone in an unsupervised laboratory?

because if you are injured there would be no one to give you assistence.

Preparing for the Laboratory

Prepare in advance for each exercise assigned. The laboratory instructor will determine whether you are to work independently, or with a partner or group. Study the entire exercise before entering the laboratory so you will know what to expect, and what data must be recorded for the written report you will make. Thought Questions, on removable report forms, are found at the end of each exercise. These may be assigned by your instructor. Satisfactorily completing these Thought Questions *before* the laboratory period will aid you in the understanding of the exercise. Valuable, usually limited laboratory time cannot be spent in this·endeavor. Performing this preliminary work will permit you to carry out your experiment efficiently.

Sometimes you may need to supplement this study by consulting your notes, the textbook, or a reference book. If certain procedures in the exercise are met for the first time, familiarize yourself with applicable information found in Part 3, "Some Laboratory Skills and Devices," and also the information in the Appendix as listed in the Contents. The following reference books are very helpful to the chemist: *Handbook of Chemistry and Physics* (CRC Press) and Lange's *Handbook of Chemistry* (McGraw-Hill Book Company).

Remember that some measurements must be extremely precise but others need only be approximate. The scientist must always keep in mind that an unnecessarily careful measurement can steal time from other work; on the other hand, the results of rough measurement can often be entirely misleading. The choice of instrument, and also skill in its operation, are important for a successful outcome.

Common Laboratory Equipment

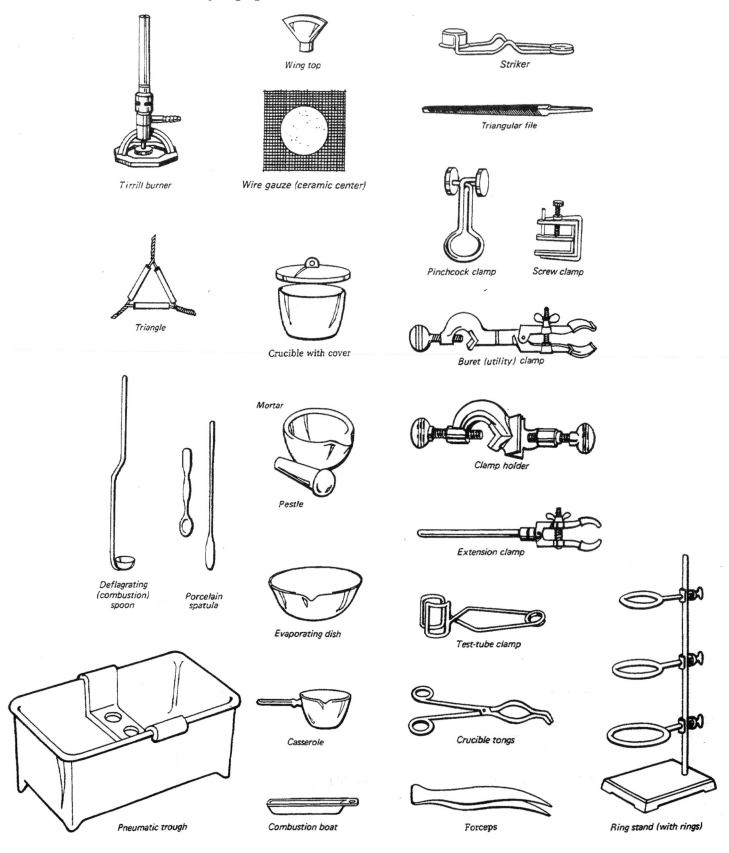

Wing top

Striker

Triangular file

Tirrill burner

Wire gauze (ceramic center)

Pinchcock clamp

Screw clamp

Triangle

Crucible with cover

Buret (utility) clamp

Mortar

Pestle

Clamp holder

Deflagrating (combustion) spoon

Porcelain spatula

Evaporating dish

Extension clamp

Test-tube clamp

Pneumatic trough

Casserole

Combustion boat

Crucible tongs

Forceps

Ring stand (with rings)

Common Laboratory Equipment (continued)

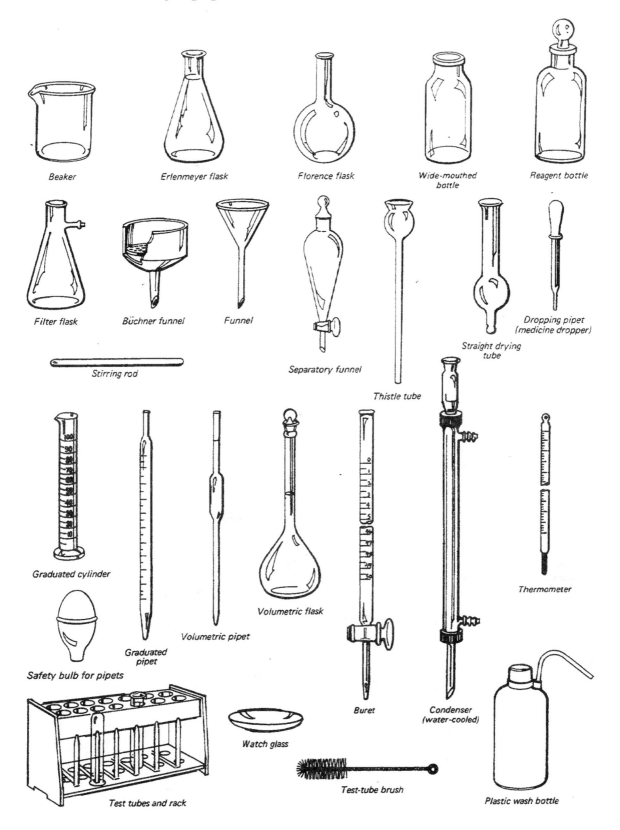

Beaker

Erlenmeyer flask

Florence flask

Wide-mouthed bottle

Reagent bottle

Filter flask

Büchner funnel

Funnel

Separatory funnel

Thistle tube

Straight drying tube

Dropping pipet (medicine dropper)

Stirring rod

Graduated cylinder

Safety bulb for pipets

Graduated pipet

Volumetric pipet

Volumetric flask

Buret

Condenser (water-cooled)

Thermometer

Test tubes and rack

Watch glass

Test-tube brush

Plastic wash bottle

D

Report Writing

A Report form, to be used as the experiment is being carried out in the laboratory, also follows each exercise. The numbers on this Report form correspond to the observation numbers embodied in the exercise. These appear as numbers in parentheses following each stratement, partial statement, or question directing your laboratory activity. For example, observation number (*1*) might follow a statement directing you to obtain a certain measurement. On the Report form for the particular exercise, you will find a corresponding observation number (*1*) preceding a space to be used for recording this measurement. Some of the observation numbers refer to a calculated result based on collected data (corresponding to previous observation numbers). Include all calculations on the Report form. Record all data at the time they are obtained. Data should be entered in ink (not pencil). No erasures are to be made; strike through incorrectly written data.

Removable Report forms have been provided. Keep a copy of each report for yourself if the original is to be turned in and not returned. Your reports are valuable when you are reviewing for examinations. Responses should be clearly and completely written. Accurate reporting of experimental results is very important in laboratory work—it is a fundamental skill that must be perfected. Calculations based on experimental data can be made with the aid of a calculator.

Part 2

Exercises

1

Measurement

OBJECTIVE

To carry out typical measurements of length, volume, and mass; to determine some conversion factors between the English and metric systems; to distinguish between accuracy and precision in the evaluation of data.

DISCUSSION

Chemistry is an experimental science. In many experiments one must obtain numerical results of typical measurements such as length, volume, mass, and density.

Measurements can be made in one system of units and then *converted* to equivalent values in another system of units. For the most part, the metric system has replaced the common (English) system of units in the scientific world, and a more recent variation of the metric system known as the SI, or International System of Units, has been in use by the U.S. Bureau of Standards since 1964. A table of units is found in the Appendix.

In each case the number obtained from a measurement *must* be accompanied by its *unit* of measurement; for example, if a certain length is measured as 1.5, one must specify whether it is 1.5 *meters,* 1.5 *inches,* or some other unit in order for this number to mean anything.

The standard unit of *length* in both metric and SI units is the meter (m). The *volume* of a body of matter is the space it occupies, and is defined in terms of the standard of length. Common metric units of volume are the liter (L) and the milliliter (mL). A liter equals the volume of a cube 1 dm (10 cm) long on each side. Thus, since the equation for the volume of a cube is $l \times w \times h$, the metric liter = 10 cm $\times$ 10 cm $\times$ 10 cm, or 1000 cm³. The unit cm³, centimeters cubed, may also be written as cc, cubic centimeters. The liter may be defined in SI units as 10^{-3} m³. The relationship exists that 1 milliliter (1 mL) = 1 cm³ = 1 cc.

The *mass* of an object is the quantity of matter that it contains. In the chemistry laboratory, mass is ordinarily measured by means of a balance whereby we "weigh" an object by adding "weights" of known mass until the balance comes to rest. Under these conditions the gravitational force acting on the object, its *weight*, is equal to that acting on the "weights." Therefore,

$$\text{Weight of object} = \text{weight of "weights"}$$

Newton's first law of motion tells us that gravitational force is directly proportional to mass:

$$\text{Weight} = k \text{ (mass)}$$

where the proportionality constant k has a fixed value at a given location. Thus, at the balance—a single given location—

$$k \text{ (mass object)} = k \text{ (mass "weights")}$$

$$\text{mass object} = \text{mass "weights"}$$

Thus the balance detects not only the equality of weight but also equality of mass. Therefore, in common terminology, the term "weight" is often used for mass of an object; that is to say, sometimes the terms "mass" and "weight" are used interchangeably. The standard unit of mass is the kilogram (kg).

Density is a property defined as the mass of a substance or object per unit volume; that is, $D = M/V$. Mass (M) is determined by weighing (see Part 3H or your instructor), and volume (V) can be determined in various ways, depending on the kind of sample. The volume of an irregularly shaped object can only be determined by the displaced volume of a lower density liquid in which it is insoluble and with which it does not react (see Fig. 1.1). Obviously the formula for the volume of a rectangular solid ($V = l \times w \times h$), or for a cylinder ($V = \pi r^2 l$), or for a sphere ($V = 4\pi r^3/3$) cannot be used.

One convenient method of conversion of units between systems, and also within each system, is the dimensional-analysis method (also referred to as the factor-label method). To use this method, a specific conversion factor is needed for each unit conversion. See the Appendix for details.

The value of data depends on how well they have been determined. Some factors involved are the skill of the operator, the precision inherently obtainable for the measuring devices used, and the validity of the method of measurement itself. The certainty of, or the confidence in, a measurement, is expressed in terms of significant figures (see Appendix). To improve confidence in experimental values, quantitative measurements are repeated a number of times. All measurements are subject to error; these errors can be separated into two different types. In multiple determinations, *systematic* (constant) errors, which are sometimes not obvious, may have occurred, such as incorrect calibrations or interfering side reactions. These errors occur in one determination after another with about the same magnitude. Usually, two or more different methods of measurements applied to the same determination will tend to minimize systematic errors. *Random* (experimental or accidental) errors may also have been made, such as human errors or those resulting from changing conditions. It can be shown mathematically that the frequency and magnitude of random errors in a series of measurements will fall on a bell-shaped (Gaussian) probability curve. That is, small errors occur more frequently than large errors, and errors on the high side occur with the same frequency as those on the low side.

When one piece of data in a large set appears to be inconsistent with the rest of the measurements, a statistical test, known as the Q test (consult a textbook on quantitative analysis) may be applied to decide whether to retain the value or to discard it as unreliable. Usually the repetition of a questionable experiment is adequate.

The terms *accuracy* and *precision* enter here. Data are accurate if the mean (average) value agrees well *with the "true" or accepted value*, implying a low degree of systematic errors. Data are precise if several determinations of the same kind agree

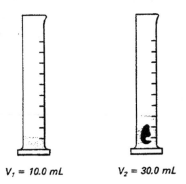

$V_1 = 10.0\ mL$ $V_2 = 30.0\ mL$

Volume of substance = 30.0 mL − 10.0 mL = 20.0 mL

Figure 1.1 Determination of the volume of an irregularly shaped object.

well *with each other*, implying a low degree of random errors. It is possible to have a result with high precision and low accuracy, or vice versa, as well as one with both high precision and accuracy, or neither.

A simplified mathematical treatment of errors, which enables one to distinguish between the accuracy and precision of measurements, is found in the Appendix.

PROCEDURE

In this exercise you will become acquainted with the technique of measuring length, volume, mass, and density. You will see how determined values can be converted from one system of measurement to another. You will also determine three different values of density for the same known substance and evaluate your experimental results as to accuracy.

1. Length

Measure the width of your desk top with a meter stick and express your answer in centimeters (cm) (*1*).

Now, using the reverse side of the meter stick, which is in English units, repeat this measurement and express your answer in inches (*2*).

From these data, calculate how many centimeters equal 1 inch (*3*).

2. Volume

Obtain a 1-qt container (or 32-oz bottle); fill it with water and measure this water, using a 100-mL graduated cylinder (see p. ix, Common Laboratory Equipment). Read the volume of each increment to the nearest milliliter at the bottom of the meniscus (see Part 3, Fig G.3).

Add the increments to get the total volume. Using this method of measuring, how many milliliters equal 1 quart (*4*)?

Referring to the table of units in the Appendix, how many milliliters actually equal 1 quart (*5*)?

Calculate the percent accuracy of your result (*6*).

Suggest a better method of carrying out this measurement (*7*).

Use the dimensional-analysis (factor-label) method to determine how many milliliters equal 1 fluid ounce (32 fluid oz = 1 qt) (*8*).

3. Mass

Obtain an object that has a known mass expressed as a fraction of a pound. Record the number or letter of this object and its weight as a fraction of a pound (*9*).

Next, change this fraction to the corresponding decimal fraction by dividing the denominator into the numerator (*10*).

Now determine the mass of this same object in grams by weighing it on a balance to the nearest 0.1 g (*11*). See Part 3H for details on weighing, or ask your instructor for help, if necessary.

Calculate how many grams equal 1 pound; use the correct number of significant figures in your answer (*12*). See Appendix for a review of significant figures.

4. Density

Obtain three different samples of the same metal or other substance supplied for this exercise. These samples are to be returned to the supply table after the exercise is completed.

Weigh each of these samples on a hanging-pan or top-loading balance to the nearest 0.01 g; remember which mass corresponds to which sample. Record these masses, all subsequent data, and calculations of results for this part in the table in the report form (*13*).

Place about 6 mL of water in a clean, 10-mL graduated cylinder. Read this volume to the nearest 0.05 mL, at the bottom of the meniscus. Record this value in your table.

Now carefully slide the first of the weighed samples into the cylinder and disperse any air bubbles. Read and record the new volume. The difference in the two volumes is the volume of the sample (volume displacement) (see Fig. 1.1).

Repeat this procedure for the remaining two samples, recording the data in the same manner.

Now calculate the density of each sample, and record, using the proper number of significant figures.

Calculate the mean (average) density of the sample used (*14*).

Find out from your instructor or some reference text the "true" density of the sample you used (*15*), and then calculate the percent accuracy of your result (*16*).

2

Identifying a Substance

OBJECTIVE

To become familiar with the role of *properties*, that is, characteristics that enable us to distinguish one substance from another, in the identification of substances.

DISCUSSION

A *substance* is any variety of matter, all specimens of which have identical properties, that is, characteristics or qualities, under the same physical conditions, and which also have the same composition. A substance is either an element or a compound. A substance may be identified on the basis of *chemical properties*, that is, properties that are observed when the substance is converted by chemical change to one or more different substances. More commonly, substances are identified by measuring their *physical properties*, such as solubility, density, melting point, boiling point, color, crystalline form, hardness, and physical state (solid, liquid, or gas). Although changes in physical conditions, such as temperature and pressure, may modify a substance's physical properties, measuring these properties does not change the chemical identity of the substance.

The *solubility* of a substance (solute) in a pure solvent at a given temperature is a physical property that is valuable in its identification. Solubility of a solute can be expressed in *quantitative* terms that give the concentration of the solution—for example, grams of solute soluble in 100 g of solvent at a definite temperature, or moles of solute per liter. In this exercise, however, we will express relative solubilities *qualitatively* by the terms *soluble, slightly soluble,* and *insoluble.*

The solubility (*miscibility*) of a liquid substance in a pure solvent is a physical property that is valuable in its identification.

Another physical property of a substance is *density,* defined as mass per unit volume ($D = M/V$). Pure substances can often be identified by determining their densities, since it is rare that any two substances have identical densities. For a more complete discussion of density, see Exercise 1, page 20.

At a fixed pressure, the temperature at which the liquid and solid phases (states) of a substance are in equilibrium is known as the *melting point* of the solid.

When bubbles of vapor form within a liquid, rise freely to the surface, and burst, the liquid is said to boil. The normal *boiling point* of a liquid is that temperature at which its vapor pressure becomes exactly equal to the standard atmospheric pressure of 760 mm Hg.

PROCEDURE

In this exercise you will be given an unknown substance (a compound) which will be either a solid or a liquid. Follow the directions in either Section 1 or 2, accordingly, and then proceed to Section 3. Fill out the corresponding sections in the Report form. You will measure the following physical properties of your unknown: solubility in two different solvents—water and ethanol; density; and melting point (if a solid) or boiling point (if a liquid). Identification of your unknown will be made by matching your set of measured properties with known properties of a compound found in the table of physical properties in this exercise.

1. Properties of a Solid Unknown

Obtain about 8 g of an unknown solid in a clean, dry container. Record its number or letter (*1*).

A. Solubility. Set up two test tubes in a rack. Add 2 mL of water to the first, and 2 mL of ethanol to the second. Now add several small crystals of your unknown to each test tube, shake each tube back and forth for a few minutes, and observe whether the crystals dissolve completely, partially, or not at all. Record as "s" for soluble, "sl s" for slightly soluble, and "i" for insoluble (*2*).

B. Density. Weigh to nearest 0.01 g about 6 g (or less, if your unknown is unusually bulky—see instructor) of your unknown solid (*3*).

Half fill a 10-mL graduated cylinder with the liquid *in which the unknown was found to be insoluble* in Part A, and read the volume to the nearest 0.05 mL (*4*). If your unknown was found to be soluble in both solvents, then this density determination cannot be carried out. Why not (*5*)?

Add the weighed solid to the liquid, being careful not to lose any material. Read the new volume to the nearest 0.05 mL (*6*). The difference in the two volumes represents the volume of the solid (*7*).

Calculate the density of the solid in grams per milliliter (*8*).

C. Melting Point. Set up a melting-point apparatus as shown in Part 3, Fig. K.1. Use water in the beaker. Obtain a capillary tube. (If your instructor directs you to make them, refer to Part 3C, page 388).

Push the open end of the capillary tube into a small amount of your unknown. Tap the bottom of the tube on your desk top to pack the solid to a depth of about 2 mm.

Attach the tube to the thermometer as shown in the diagram (Fig. K.1 in Part 3) by means of a rubber band or a thin section of proper-sized rubber tubing. Be careful not to get any water into the tube.

Heat the water in the beaker slowly (1°C per minute). Note the temperature at which the last crystal *just* melts, and record this as the melting point of your unknown (*9*). If you are not certain of your result, repeat the procedure with a new capillary tube and sample.

Discard the used capillary tubes in the solids waste jar.

2. Properties of a Liquid Unknown

Obtain about 20 mL of an unknown liquid in a clean, dry container. Record its number or letter (*10*).

A. Solubility (Miscibility). Set up two test tubes in a rack. Add 2 mL of water to the first, and 2 mL of ethanol to the second. Now add 2 mL of your unknown to each tube, shake briefly, and observe whether the two liquids mix completely, partially, or not at all. If the liquids mix completely (are miscible), there will be the end appearance of only one liquid present; if the liquids do not mix at all (are immiscible; that is, "insoluble"), a line of separation will still be visible after shaking the tubes and then allowing them to stand for a minute or so. The unknown liquid may be "partially soluble," as judged by the amount remaining separate after shaking as compared to the amount initially added. Record as "s" for soluble, "sl s" for slightly soluble, or "i" for insoluble (*11*).

B. Density. Carefully weigh a clean, dry 10-mL beaker to the nearest 0.01 g (*12*). If 10-mL beakers are not available, use the smallest size possible.

Deliver exactly 10 mL of the unknown liquid to the beaker, using a *clean* and *dry* pipet with a safety-filling device (see Part 3G, page 391 for details) if available; if not, then use a 10-mL graduated cylinder.

Weigh the beaker and contents (*13*).

What is the weight (mass) of the liquid (*14*)?

From the data obtained, calculate the density of the liquid in grams per milliliter (*15*).

C. Boiling Point. Set up a boiling-point apparatus as shown in Part 3, Fig. J.1. Fill the test tube to a depth of about 3 cm with the unknown liquid. Add a very small boiling chip to the liquid to ensure even boiling. Heat the water very gradually and watch for changes in the test tube.

Record the temperature at which the liquid in the test tube boils freely (16).

3. Identifying the Unknown Substance

Your unknown is one of the substances given in the following table of physical properties. Compare the set of properties you have determined for your unknown with those listed in the table and identify the unknown (17).

PHYSICAL PROPERTIES OF MISCELLANEOUS SUBSTANCES

Substance	Density (g/mL)	Melting point (°C)	Boiling point (°C)	Solubility	
				Water	Ethanol
Borax	1.73	75	dec 200	sl s	i
Bromoethane	1.43	−118	38	i	s
Cadmium nitrate · 4 H₂O	2.46	59	132	s	s
Calcium carbonate (chalk)	2.93	dec 825	dec	i	i
Calcium hydroxide (slaked lime)	2.2	dec 580	dec	sl s	sl s
Calcium nitrate · 4 H₂O	1.82	43	dec	s	s
Cyclohexane	0.78	6.5	81	i	s
p-Dichlorobenzene	1.46	53	174	i	s
Diphenyl (Biphenyl)	0.99	70	255	i	s
Diphenylmethane	1.00	27	265	i	s
Ethanol	0.79	−112	78	s	s
Heptane	0.68	−91	98	i	sl s
Hexane (component of gasoline)	0.66	−94	69	i	s
Iodomethane	2.28	−66	42	sl s	s
Lauric acid (found in coconut oil)	0.88	48	225	i	s
Methanol	0.79	−98	65	s	s
Naphthalene	1.15	80	218	i	sl s
2-Propanol	0.79	−86	83	s	s
Propanone	0.79	−95	56	s	s
Stearic acid (found in animal fats)	0.85	70	291	i	sl s
Thymol	0.97	52	232	sl s	s
p-Toluidine	0.97	45	200	sl s	s
Trichloromethane	1.49	−63.5	61	i	s
Zinc nitrate · 6 H₂O	2.06	36	dec 105	s	s

Note: Symbols used in this table are s = soluble, sl s = slightly soluble, i = insoluble, dec = decomposes. Values are rounded off.

3

Chemical and Physical Changes

OBJECTIVE

To become familiar with the principles used in differentiating between chemical and physical changes.

DISCUSSION

A *chemical change* is the alteration of a substance into one or more than one entirely different substance. Since each substance possesses its own set of properties, chemical·changes are always accompanied by the appearance of new sets of properties corresponding to formation of new substances. Thus, when an oxide of mercury is heated, the red powder gradually disappears and a silvery liquid condenses on the cooler parts of the vessel. When iron rusts, the corrosion of iron metal contains oxygen as well as iron, and is therefore a different substance from iron metal and has different properties. The energy content of substances is a property, and the difference in energy content of reactants and products is evidenced as energy lost or gained during the reaction. Although energy may be gained or lost in several forms, thermal energy is by far the most common form in chemical reactions. In *physical changes*, energy gain or loss also occurs, but since no new substances are produced, the other properties remain the same. Only physical properties are altered during physical changes. Examples of physical changes are the melting of ice, the freezing of water, and the conversion of water to steam. In each of these cases there is no change in the chemical composition of the substance involved, namely, water.

PROCEDURE

Your instructor will inform you as to which sections of this exercise you are to complete. In Section 1 of this exercise you will carry out several experiments and classify the results of each as either a chemical or physical change by comparing the properties of the substances before and after. In Section 2 you will observe and tabulate the properties of two particular elements, iodine and antimony. In Section 3, you will carry out two different experiments involving iodine and antimony, and again classify the results as either a chemical or physical change, based on the properties of each element observed in Section 2.

1. Some Examples of Chemical and Physical Changes

Observe carefully any changes that occur in procedures **A** through **G**. Describe the changes in the Report form, and on this basis classify each result as a chemical or physical change. If directions on heating are needed, see Part 3B.

A. Heat strongly a few crystals of copper(II) nitrate in an uncovered crucible (*1*). **Caution: Carry out this procedure in the hood.**

B. Hold a 1-inch strip of magnesium ribbon in a flame by means of tongs until a change is observed (*2*). **Caution: Shield eyes by looking away.**

C. Hold a nichrome wire in the hottest part (see Fig. B.2 in Part 3) of the flame for a few seconds; remove and allow to cool (3).

D. Hold a piece of copper metal sheet in the hottest part of the flame for a few minutes; remove and allow to cool (4).

E. Add 2 mL of 0.1 M barium chloride solution to a like amount of 0.1 M sodium sulfate solution contained in a test tube (5).

F. Add 2 mL of 1 M hydrochloric acid to a small amount of calcium carbonate contained in a test tube and shake gently (6).

G. Mix 3 mL of 0.1 M sodium iodide solution with 3 mL of clorox bleach in a test tube (7).

2. Identification of Iodine and Antimony

Reminder: Goggles must be worn. Avoid skin contact and inhalation of vapors.

Examine a small quantity of iodine and note its color, odor, and crystalline form (8).

Test the solubility of iodine in water and in dichloromethane by adding a small crystal to 1 mL of each of the liquids in separate glass test tubes, and shaking gently. Also note the color of each solution (9).

Evaporate all the dichloromethane (**Caution: Use the hood.**) from the iodine solution by heating gently. What is the residue (10)?

Heat the residue gently. What happens (11)?

Using clean test tubes, examine a *small* sample of powdered antimony as you examined iodine, *also evaporating the dichloromethane in the hood.* List the properties of antimony that you have observed (12).

3. Interaction of Iodine and Antimony

Note: Under no circumstances may class-size amounts of this mixture be prepared by the instructor; an explosion may occur.

Reminder: Goggles must be worn for procedures A and B.

Using a clean, dry watch glass as a mortar, and the end of a test tube as a pestle, grind two large crystals of iodine. Add a quantity of powdered antimony about the size of a match head, but do not grind the iodine and antimony together. Using a spatula or a glass stirring rod, *gently mix* and then divide the mixture into two portions; keep one portion on the watch glass for procedure A, and transfer the other portion (use a clean, dry spatula) to a dry 200 × 25-mm Pyrex test tube for procedure B.

A. Add 2 mL of dichloromethane to the portion of the mixture on the watch glass and mix thoroughly.

Set up a filtration assembly as shown in Fig. D.3 in Part 3. Filter the mixture into a test tube and wash the residue on the filter paper with about 20 drops of dichloromethane (use a medicine dropper).

Spread the paper (with the residue on it) out to dry. Is the residue a different substance (13)? Base this observation on the properties of iodine and antimony determined in Section 2.

Heat the test tube (**Caution: Use the hood.**) containing the filtrate until all the liquid has evaporated; then continue heating for a short time. Is the residue in the test tube a different substance (14)? Again, base this observation on the properties of iodine and antimony determined in Section 2.

What is your conclusion about the change that occurred in procedure A? Was it a physical or chemical change (15)? Give reasons for your decision (16).

B. Using the second portion of the mixture of antimony and iodine in the 200 × 25-mm Pyrex test tube, add 4 mL of dichloromethane as the solvent. Insert a thistle tube (see page 13, Common Laboratory Equipment) into a one-hole stopper that fits the test tube. The thistle tube should extend about one-half inch below the bottom of the stopper.

Clamp the test tube with the stopper and thistle tube on a ring stand, and heat the test tube *very gently over a low flame* for 20 minutes. Do not heat so strongly that more than half of the liquid is in the upper part of the apparatus at any one time. If this occurs, allow the apparatus to cool somewhat before applying any more heat. Note and describe any changes that occur during the heating of the mixture (17).

Discontinue the heating, allow the solid material to settle, and decant (pour off) the hot supernatant liquid into a clean, dry test tube, being careful not to transfer any of the solid. Describe any changes that occur as the liquid cools (*18*).

Evaporate the liquid to half its volume and again allow it to cool. If crystals form, filter them off and heat them gently in an evaporating dish. If no crystals form, evaporate more of the liquid and again allow it to cool. Note the result (*19*).

Is the product of the change in procedure B iodine, antimony, or an entirely different substance (*20*)? Was it a physical or chemical change (*21*)? Give reasons for your answer (*22*).

NAME _____ SECTION _____ DATE _____

1. Some Examples of Chemical and Physical Changes

Description of change	Kind of change
(1)	
(2)	
(3)	
(4)	
(5)	
(6)	
(7)	

2. Identification of Iodine and Antimony

(8)–(11) Properties of iodine:

(12) Properties of antimony:

3. Interaction of Iodine and Antimony

A. DESCRIPTION OF CHANGE

(13)

(14)

(15), (16) Kind of change:

B. DESCRIPTION OF CHANGE

(17)–(19)

(20) Product of change:

(21), (22) Kind of change:

NAME _____ SECTION _____ DATE _____

Answer the following questions before beginning the exercise:

1. What are the important points to be observed in carrying out gravity filtration through paper (see Fig. D.3, Part 3)?

2. Generally speaking, what characterizes a chemical change?

3. Write equations for any chemical changes that occur in Section 1 of this exercise. Consult your textbook or other reference where necessary.

4. What would be the formula for the product of a chemical combination of antimony and iodine (refer to a Periodic Chart)?

4

Separating Components of Mixtures

OBJECTIVE

To become familiar with the techniques of filtration, extraction, distillation, and sublimation as means of separating the components of mixtures.

DISCUSSION

When two or more substances that do not react chemically are combined, a *mixture* results. Thus, a *mixture* is a sample of matter containing more than one substance; the relative proportions of its components may vary widely. If the mixture consists of only one phase (that is, it is uniform throughout), it is referred to as *homogeneous*, or a solution. Air, for example, is a homogeneous mixture of several different gases including nitrogen, oxygen, argon, water vapor, and carbon dioxide. A mixture of a small amount of sugar and water is a solution. If the mixture consists of more than one phase (that is, it is nonuniform throughout), it is referred to as *heterogeneous*. Most of the rocks and minerals in the earth's crust are complex heterogeneous mixtures of many different chemical substances. Each of the substances in a mixture retains its chemical integrity, and mixtures are separable by physical means. Separating and identifying the components of mixtures is one of the most important problems the chemist is called on to solve. The methods chosen depend on the type of mixture, but in all cases advantage is taken of the fact that the different components have different properties.

PROCEDURE

In this exercise you will devise and carry out a method for separating the components of a prepared mixture, in terms of the physical properties of the components and a choice of the techniques listed. Note, however, that many more separation techniques exist.

Obtain a 5-g sample of the mixture that has been prepared for this exercise; it is advisable to shake the storage bottle before taking the sample. Weigh the sample to 0.01 g (*1*). Copy its composition from the label (*2*). Obtain a list of the physical properties of the components of the mixture from your instructor, a handbook, or a handout, and record these data (*3*).

Outline a workable scheme in terms of the following techniques, with the aid of the list of properties obtained, for carrying out the separation of the substances present (*4*). Be sure to check this proposed scheme with your instructor before proceeding.

Carry out the separation and report the mass of each substance recovered (*5*).

Compare your results with the composition given and compute the error in recovery of each component (*6*). Try to account for any large discrepancies and suggest possible improvements in your procedure (*7*).

Techniques for Separating Mixtures
(Demonstration Setups Are Recommended)

A. Filtration. Gravity filtration, described in Part 3D, is a method for separating a solid and a liquid. The liquid, and any substance soluble in the liquid, passes through the filter paper and is termed the *filtrate*. The solid (or insoluble substance) remains on the filter paper. The soluble substance in the filtrate can be separated by evaporation of the solvent. Evaporation may be accomplished by gently heating the filtrate contained in an evaporating dish over a Bunsen flame, or on a hot plate if the solvent is flammable. (Check with your instructor.)

B. Extraction. Extraction is a method for separating two immiscible (mutually insoluble) liquids, as well as any solute that either may contain. When a solute is added to two immiscible solvents, it will dissolve in each, but usually unequally. This behavior is often of practical importance. Iodine, for example, is soluble in both water and dichloromethane, whereas water and dichloromethane are nearly immiscible and form separate layers when in contact. When an aqueous (water) solution of iodine is shaken with an equal volume of dichloromethane, the greater part of the iodine will leave the water and go into the dichloromethane, in which it is more soluble. This process is known as *extraction*. If the aqueous layer is extracted a second time with a fresh portion of dichloromethane, the concentration of iodine in the water solution can be reduced further. Figure 4.1 illustrates this technique.

C. Distillation. Distillation is a method for separating substances of differing volatility. When a

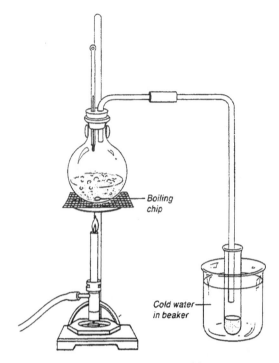

Figure 4.2   A simple distillation assembly using a test tube immersed in cold water as a condenser.

liquid is heated in a distillation apparatus, the liquid is converted to vapor. The vapor, on passing through the condenser, is reconverted to liquid, and the liquid flows into the receiving vessel. Substances differing sufficiently in volatility can be separated in this way. Distillation illustrates the principle that adding heat to a liquid favors evaporation, an endothermic change of state, and removing heat favors condensation, an exothermic change of state.

Figure 4.1   A capillary-tipped medicine dropper being used to separate two immiscible liquids.

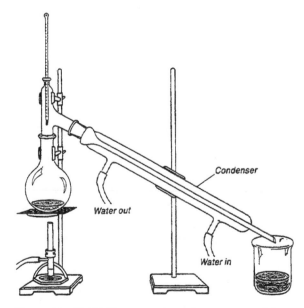

Figure 4.3   Distillation apparatus using a commercial water-cooled condenser.

A simple distillation apparatus, using an air-cooled condenser, can be set up as shown in Fig. 4.2, using a 250-mL Florence flask. The Florence flask is clamped to the ring stand as shown. The applications of this setup are limited (see instructor). If available, a distillation setup with a water-cooled condenser, similar to the one shown in Fig. 4.3, is more versatile.

D. Sublimation. Sublimation is a method for separating substances that pass directly from the solid state to the gaseous state upon heating, without passing through the liquid state. Conversely, vapors can condense directly to the solid state. For this experiment, the material that contains a substance that sublimes is placed in a 100-mL beaker supported high above a small flame. The beaker is covered with a watch glass containing crushed ice. The sublimate condenses on the bottom of the watch glass.

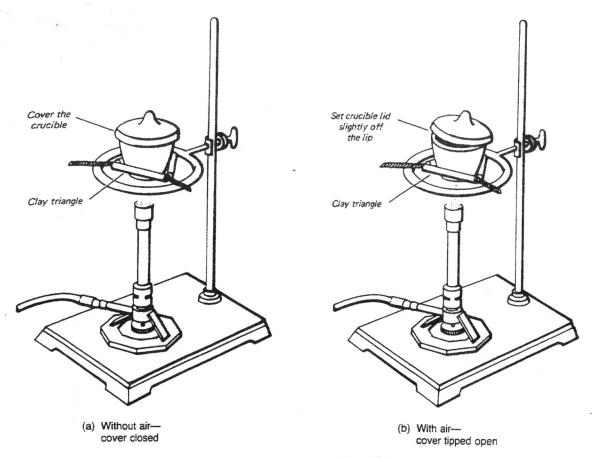

(a) Without air—
    cover closed

(b) With air—
    cover tipped open

Figure 5.1 Heating a crucible and lid.

number ratio. In any case, however, you must use the data obtained in your calculations.

To avoid breakage of the crucible or its lid during the exercise, practice handling the empty crucible using crucible tongs *before* starting the exercise. In particular, practice displacing the lid slightly; this maneuver requires care and skill (see Fig. 5.1).

1. Formula of an Oxide of Magnesium

Support a clean crucible *without its cover* on a clay triangle and dry by heating for 5 minutes with a Bunsen flame as shown in Fig. 5.1. Turn off the flame and allow the crucible to cool to room temperature. Handle the crucible with crucible tongs.

Weigh the *cooled* crucible (without its cover) to 0.01 g (*1*). Place a 0.3-g sample of magnesium turnings or magnesium ribbon in the bottom of the weighed crucible. Magnesium turnings are preferable, but if magnesium ribbon is used, it must be crumpled or *loosely* rolled. Reweigh the crucible plus the magnesium to 0.01 g (*2*).

Cover the crucible with its contents and replace on the triangle; heat with the cover in place [see Fig. 5.1(a)], first gently and then more strongly for about 5 minutes.

Now, using crucible tongs, *carefully* displace the lid slightly to allow more air in [see Fig. 5.1(b)], and heat strongly for 10 minutes. Use the hottest part of the flame, but be careful not to ignite the magnesium; that is, do not allow the flame to lap over the top of the crucible.

Remove the flame, *cool somewhat,* and then add 2 to 3 drops of water to convert any magnesium nitride (a side reaction that may occur) to the hydroxide.

Again heat, gently at first and then strongly, for 5 minutes, with the lid slightly displaced.

If magnesium ribbon has been used, break up any lumps with a glass rod, but be sure not to remove.

any material with the rod. If the residue is not gray or white, cool, add water again, and reheat.

Allow the covered crucible to cool while on the clay triangle to room temperature. Remove the cover. Weigh the crucible and its contents (*3*).

Record your data and calculations in the Report form.

Repeat the determination, using a clean, dry crucible.

2. Formula of a Sulfide of Copper

Support a clean crucible *without its cover* on a clay triangle and dry by heating for 5 minutes with a Bunsen flame as shown in Fig. 5.1. Turn off the flame and allow the crucible to cool to room temperature. Handle the crucible with crucible tongs.

Weigh the *cooled* crucible (without its cover) to 0.01 g (*4*). Place about 0.4–0.6 g of copper wire (rolled into a spiral) or copper turnings in the bottom of the crucible and reweigh to 0.01 g (*5*).

Add sufficient powdered sulfur (about 1 g) to *cover* the copper; keep the sulfur from getting all over. Close the crucible with the cover, and replace it on the triangle. *Move the setup into the hood.*

Gently heat the crucible until the blue flames of burning sulfur vapor are no longer seen at the edge of the lid. Do not remove the cover. Finally heat the crucible with the hottest part of the flame to a dull red for 5 minutes. Allow the *covered* crucible to cool to room temperature on the triangle.

When cool, remove the cover, and weigh the crucible and its contents to 0.01 g (*6*).

Add a small quantity of sulfur to the crucible, cover it, and repeat the process of heating (*in the hood*), cooling, and weighing (*7*). If the last two weighings agree to within 0.1 g, the chemical reaction between the copper and sulfur is complete. If not, repeat the process.

Record your data and calculations in the Report form.

NAME _____ SECTION _____ DATE _____

2. Formula of a Sulfide of Copper

(4) Weight of a crucible .... _____ g

(5) Weight of crucible + copper .... _____ g

Weight of copper (5) − (4) .... _____ g

Moles of copper (show calculation) .... _____

(6) Weight of crucible + copper sulfide after first heating .... _____ g

(7) Weight of crucible + copper sulfide after second (or final)

heating .... _____ g

Note: (6) and (7) should agree to within 0.1 g. If they do not, see instructor before proceeding.

Weight of sulfur (7) − (5) .... _____ g

Moles of sulfur (show calculation) .... _____

Using these data, complete the following table:

Element	Mass of Element	Relative Number of Moles	Divide by the Smaller Number	Smallest Integral Number of Moles
Copper				
Sulfur				

Empirical (simplest) formula for copper sulfide _____

6

Water of Hydration

OBJECTIVE

To determine the percent of water in a hydrated salt; to determine the formula for the hydrate, knowing the percent of water and given the formula for the anhydrate.

DISCUSSION

Salts are compounds composed of positive ions (cations) and negative ions (anions), held together in the solid state by strong ionic bonds. When certain salt solutions are allowed to evaporate, some water molecules remain as part of the crystalline salt left after evaporation is complete. These salts and water are combined in definite proportions, and are known as *hydrates*. Two examples are $CaCl_2 \cdot 2H_2O$ (calcium chloride dihydrate), and $Na_2CO_3 \cdot 10H_2O$ (sodium carbonate decahydrate). The centered dot between the formula for the salt and the formula for the water indicates a bond between the salt and the water, which is known as *water of hydration*. When hydrates are heated, this bond, which is weaker than the bonds in the salt itself, usually is broken and the water is driven off, leaving an anhydrous salt, the *anhydrate*. For example,

$$CaCl_2 \cdot 2H_2O \xrightarrow{\Delta} CaCl_2 + 2H_2O$$

The formula mass of $CaCl_2 \cdot 2H_2O$ is 147 g; it contains 27.3% Ca, 48.3% Cl, and 24.5% H_2O. When anhydrous $CaCl_2$ is exposed to water, the foregoing equation is reversed, and the hydrate is reformed.

PROCEDURE

You will be given a hydrated salt, and the formula for the corresponding anhydrous salt. From this information you will determine the percent of water in the hydrate, and the formula for the hydrate.

Obtain about 3 g of a hydrate. Record the formula of the corresponding salt (1).

Obtain a clean crucible and its matched cover. Practice, with crucible tongs, successfully tipping the cover slightly, as shown in Fig. 5.1(b). Support the crucible and its tipped cover on a clay triangle and dry by heating for 5 minutes with a Bunsen flame as shown in Fig. 5.1(b). Turn off the flame and allow the crucible and cover to cool to room temperature (about 10 minutes). Accurate

weighing cannot be made if not cooled to room temperature. Handle the crucible and cover with crucible tongs from this point on, so that moisture from the hands will not be transferred.

Weigh the cooled crucible and cover to 0.01 g (2). Place about a 2-g sample of your hydrate in the crucible and weigh, with the cover, to 0.01 g (3).

Place the covered crucible on the clay triangle, tip the cover slightly to allow the water vapor to escape, and *very gently* heat the crucible for about 5 minutes. Then heat strongly for another 15 minutes. The burner flame should be so adjusted that the crucible bottom becomes dull red during this period of heating. Then remove the flame, *close the cover*, and cool to room temperature (about 10 minutes). Weigh to 0.01 g (4).

Repeat the heating at maximum temperature, with cover tipped, for an additional 6 minutes, cool with cover closed, and reweigh (5).

If the decrease in weight between these two weighings differs by more than 0.1 g, repeat once more (6). It is necessary for a constant weight to be obtained, to ensure that all the water of hydration has been driven off. Record the final weight (7).

Calculate the percent of water in your hydrate (8).

Now calculate the gram formula mass of your anhydrous salt by adding the gram atomic masses in the formula of your sample (9). Determine the number of moles of anhydrous salt in the amount of sample used (10). Determine the number of moles of water of hydration driven off from the sample (11).

The ratio of the moles of anhydrous salt to the number of moles of water can be simplified to whole numbers by dividing each ratio by the smaller of the two numbers (12).

Using this ratio of whole numbers, write the formula of your hydrate as follows:

anhydrate (expressed as its formula) $\cdot$ $x\,H_2O$ (13),

where x is the number of moles of water of hydration per mole of the anhydrate.

Determining Atomic Mass from Specific Heat

OBJECTIVE

To become familiar with a method involving calorimetry for determining atomic mass that has both theoretical and historical significance.

DISCUSSION

The discovery by Dulong and Petit in 1819 that the product of the specific heat and atomic mass is very nearly the same for many solid elements (about 6.4) was a milestone in the development of the atomic theory. This law has been found to be approximately valid for all solid elements having atomic masses greater than 40 and for metallic elements. It does not hold (at room temperature) for such elements as carbon, silicon, phosphorus, or sulfur. Before Dulong and Petit made this discovery, atomic masses could only be conjectured from experimentally derived equivalent weights by certain dubious assumptions. With this law, decisions could be made about which multiple of the equivalent weight of an element represented its atomic mass. Although the law of Dulong and Petit was proposed originally as an empirical rule, it has been shown in recent years to have a basic explanation in thermodynamics and atomic structure theory. The classic derivation and its prediction of a heat capacity independent of temperature is now replaced by quantum-mechanical treatments (the Einstein and the Debye theories) of the vibrational frequencies assigned to a crystalline solid containing an Avogadro's number of single particles, and the Debye low-temperature limiting expression. For further information on this subject, consult a textbook on physical chemistry.

PROCEDURE

In this exercise you will determine the atomic mass of an unknown metal, using the principle of Dulong and Petit:

$$\text{Atomic mass} = \frac{6.4}{\text{specific heat}}$$

Specific heat is defined as the number of calories of heat required to raise the temperature of one gram of a substance 1°C. The units of specific heat in this instance are cal/g · °C.

Construct a calorimeter as follows: Cut a piece of cardboard to be a cover for a 600-mL beaker, or a foam plastic cup. In the center of the cardboard, cut a circular hole large enough to support an aluminum tumbler (volume about 150 mL) by its rim, or one fashioned from flexible scrap aluminum. Next, prepare a cardboard cover for the tumbler, and cut a hole large enough to accommodate *snugly* a cork holding a thermometer. Assemble the calorimeter (see Fig. 7.1) and adjust the thermometer so that its bulb is 1.5 cm from the bottom of the tumbler. If a beaker is used, wrap the beaker in a towel for added insulation. Have the instructor check your setup.

Obtain about 100 g of a metal from the instructor and place it in a clean, dry 200 × 25-mm test tube

69

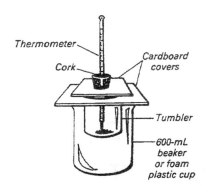

Figure 7.1 Assembled calorimeter for determining atomic mass from specific heat.

that has been carefully weighed (*1*). Weigh test tube and contents (*2*), and record the weight of the sample (*3*). Cork the test tube lightly and place it in boiling water for at least 20 minutes. Record the water temperature (*4*). While the sample is being heated, deliver *exactly* 50 mL of water, preferably from a volumetric pipet, into the tumbler and replace the cover.

When 20 minutes have elapsed, read the temperature of the water in the tumbler, estimating to 0.1°C (*5*). Now transfer the metal quickly, without splashing, into the tumbler. Replace the cover, swirl the contents briefly, and read the *highest* temperature reached (*6*).

Assuming heat loss to the surroundings to be negligible, the heat lost by the metal (mass × specific heat × temperature change) equals the heat gained by the water and the calorimeter. An approximate value for the heat the calorimeter gained can be obtained from its weight and specific heat. Dry and weigh the aluminum tumbler. The specific heat of aluminum is 0.214 cal/g · °C. The heat the calorimeter gained can be estimated from these data and the temperature rise. Record the temperature change of the metal (*7*) and water and tumbler (*8*). Compute *H*, the specific heat of the metal sample (*9*). Dry and return the metal to your instructor or supply table.

Using the law of Dulong and Petit, compute the atomic mass of the metal and record it in the Report form (*10*). What is the metal (*11*)? What is your percentage error (*12*)?

List some of the main sources of error in this experiment and indicate for each whether it would tend to make the calculated atomic mass high or low (*13*).

Molecular Mass Determination by Vapor Density Method

OBJECTIVE

To become familiar with a fundamental method for determining molecular mass, an important quantity in chemistry.

DISCUSSION

Avogadro's hypothesis, that equal volumes of gases at the same temperature and pressure contain equal numbers of molecules, provides a direct method for obtaining the relative masses of molecules of gases and vaporizable liquids and solids. When vapor density (mass per volume) is introduced into the ideal gas equation $PV = nRT$, the value of molecular mass determined is sufficiently accurate for general purposes. The pressure, volume, and temperature of the vapor of the liquid to be studied is measured. Using an appropriate value of the ideal gas constant R, one can calculate the molecular mass of the vapor since

$$n = \frac{\text{mass}}{\text{molecular mass}} = \frac{g}{M}$$

and

$$M = \frac{gRT}{PV} = \rho \frac{RT}{P}$$

where ρ = vapor density

$$R = 0.0821 \frac{\text{L} \cdot \text{atm}}{\text{mol} \cdot \text{K}} = 62.4 \frac{\text{L} \cdot \text{mm}}{\text{mol} \cdot \text{K}} = 6.24 \times 10^4 \frac{\text{mL} \cdot \text{mm}}{\text{mol} \cdot \text{K}}$$

This method can be used for liquids that boil at a temperature lower than the boiling point of the hot liquid bath to be used—in this exercise, water.

PROCEDURE

Determine from your instructor whether you should run one or two molecular mass determinations on the known substance, 2-propanol, and whether you are also to do a determination on an unknown substance. **You should follow the directions carefully, inasmuch as they are critical in obtaining good results.**

Clean and dry a 125-mL (or 250-mL) Erlenmeyer flask by means of a hot air blast or a Bunsen flame. Allow it to cool (**Caution: Avoid burns.**) on a clean surface, such as a towel.

Obtain a small rubber band and a small square of aluminum foil. Place the square of aluminum foil over the mouth of the flask and fold the edges loosely around the bead of the flask's rim. With a pin or needle, puncture as small a hole as possible in the approximate center of the foil. The hole should be almost invisible to the naked eye. If you puncture a large hole, the experimental results will be very poor. Weigh the flask, foil, and rubber band to 0.01 g (1).

Put in the flask about 3 mL of 2-propanol (CH_3)$_2$CHOH plus a *small* crystal of iodine (to add color to the liquid). Replace the foil. Pull the foil down around the rim of the flask. Fasten the rubber band around the foil and flask, just under the rim of the flask. *Be sure it is tight. Exercise care not to tear or puncture the foil.* (If you do tear the foil, start over.)

Set up a boiling-water apparatus using a 600-mL or 800-mL beaker, as shown in Fig. 8.1. Add enough water to keep the level around the flask as high as possible. A boiling chip may be added to ensure even boiling. Heat the water to boiling. Keep a supplemental beaker with boiling water available, in order to maintain a constant level, should some water evaporate during boiling.

When the water is boiling, plunge the prepared Erlenmeyer flask into the boiling water, and clamp in place as shown in Fig. 8.1. Boil the water until no more liquid 2-propanol (colored by the iodine crystal) can be seen inside the Erlenmeyer flask. Continue boiling for another 2 minutes. Under no circumstance remove the flask from the bath, even temporarily, before this time.

Remove the flask, wipe it dry, set it on a clean towel, and allow it to cool to room temperature. Blot any droplets of water on the foil. Weigh flask and contents to 0.01 g (2).

Record the boiling point of water (3) at the prevailing barometric pressure (4). When you read the thermometer, be sure it is not touching the bottom of the beaker.

If time is available, add 3 mL more of 2-propanol and repeat the determination; record these data under Trial 2 in the Report form.

Determine the volume of the flask by filling it with water and measuring the volume of the water with a graduated cylinder (5).

Calculate the molecular mass for each determination (6).

Compute the average molecular mass (7) and the average deviation (8).

Compare your result with the correct molecular mass of 2-propanol and express your error in parts per thousand (9).

If your instructor so directs, obtain 10 mL of a liquid unknown and make two determinations of its molecular mass. Be certain the Erlenmeyer flask you use is *dry!* Record the data in the Report form.

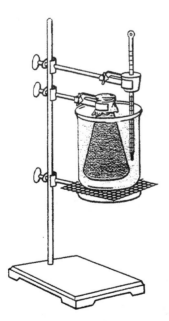

Figure 8.1 Apparatus used in determining molecular mass by the vapor density method.

9

Enthalpies of Neutralization by Solution Calorimetry

OBJECTIVE

To study the application of calorimetry to energy changes in chemical reactions.

DISCUSSION

Heat (thermal energy) absorbed or evolved in a chemical reaction can be measured by a calorimeter, insulated to restrict the heat exchange to the calorimeter and contents; the heat does not escape to the surroundings. The heat absorbed by a reaction, a quantity designated Q, equals the heat lost by the calorimeter and contents; the heat evolved by a reaction equals the heat gained by the calorimeter and contents. This relation can be expressed as $Q = -C(T_2 - T_1)$, where C is the heat capacity of the calorimeter and contents. The value C is the product of the mass of the water or solution in the calorimeter and the *specific heat* of water, corrected for the heat absorbed by the calorimeter itself. Specific heat is defined as the amount of heat needed to raise the temperature of 1 g of a substance 1 degree Celsius. The specific heat of water (and of most dilute solutions) is 1.00 cal/g · °C. Calorimetry is an application of the first law of thermodynamics to chemical and physical changes.

When a reaction is carried out at constant pressure, as it is in this exercise where the student-constructed calorimeter is open to the atmosphere, the heat flow Q is known as the change in enthalpy ΔH. The symbol Δ (Greek delta) signifies the change in some variable. The value ΔH is determined as energy per mole. From chemical thermodynamics, $\Delta H = \Delta E + P \Delta V$. At constant pressure and when volume changes are negligible, $P \Delta V = 0$, and $\Delta H = \Delta E$, the energy change in the system under consideration. When heat is absorbed, the reaction is said to be *endothermic*, and ΔH has a positive value; when heat is evolved, the reaction is said to be *exothermic*, and ΔH has a negative value.

Enthalpy changes are dealt with in terms of chemical equations, which include the physical state of each substance given in parentheses; for example,

$$H_2(g) + \frac{1}{2} O_2(g) \longrightarrow H_2O(l), \quad \Delta H_{f298.15} = -285.8 \text{ kJ mol}^{-1}$$

Thus, according to convention, this equation represents an exothermic reaction, since heat is evolved. Useful tables of standard molar enthalpies of formation, $\Delta H°_{f298.15}$ exist in which the values listed are for the formation of moles of pure substances formed from free elements in their most stable states under standard state conditions [298.15 K (25°C), 1 atm]. In the above case,

$$\Delta H^\circ_{f298.15} \text{ of } H_2O(l) = -285.8 \text{ kJ mol}^-$$

By using these tables where possible, together with experimentally determined enthalpy changes in related reactions, one can calculate the ΔH of a reaction without actually carrying out the reaction. According to Hess's law, which defines this relation, chemical equations can be treated as algebraic expressions and the corresponding enthalpy changes can be similarly treated; the result yields the enthalpy change for the desired reaction.

In other words, Hess's law may be summed up as follows: If a process can be considered to be the sum of several stepwise processes, the enthalpy change for the *total* process equals the *sum* of the enthalpy changes for the various steps. Thus for this exercise, $\Delta H_{reaction} = \Delta H_1 + \Delta H_2 + \Delta H_3$. This means that the ΔH for the reaction; namely, the enthalpy of neutralization of CH_3COOH and NH_3, can be calculated from the values determined in parts 2, 3, and 4, which represent ΔH_1, ΔH_2, and ΔH_3.

In this exercise, solutions are to be measured with volumetric pipets. If ordinary thermometers are used, readings must be estimated to 0.1°C. (Thermometers graduated in intervals of tenths of degrees are desirable and should be read to 0.03°C.)

PROCEDURE

In this exercise you will work in pairs. Each pair of students will do Section 1, Heat Capacity of the Calorimeter, and *one* of the other sections as assigned by the instructor. At the end of the laboratory period, the data from Sections 2, 3, and 4 can be correlated so that each pair of students will have the necessary data to complete Section 5, The Law of Hess.

1. Heat Capacity of the Calorimeter

Since the calorimeter itself absorbs some heat, one must establish a correction for this. Construct a calorimeter similar to the one shown in Fig. 9.1. Nest a foam plastic drinking cup of about 200-mL volume inside another. Fashion a lid from a sheet of foam plastic. Rotate the lid against the top of the cup to form a groove; a better seal is produced in this way. Include a wire stirrer as shown. With a cork borer of the proper size, make a hole in the lid just large enough to admit a thermometer. Cut a ring from a piece of rubber tubing and slip this on the thermometer to keep it from slipping too far into the calorimeter. Label this thermometer No. 1 and use it in the calorimeter.

Label a second thermometer No. 2. Compare their readings after immersing them side by side in 500 mL of tap water. Record any correction to be applied to thermometer No. 2 to make it agree with thermometer No. 1 (*1*). All readings made with thermometer No. 2 in this exercise are to be corrected, if necessary.

Place the calorimeter on an insulating surface such as cardboard. Hold it within the ring of a ring stand for stability but prevent contact with the iron by rubber tubing sleeves on the ring. Pipet 50.0 mL of water into a dry beaker and cool it to about 10°C. Heat 50.0 mL of water to about 40°C in another beaker.

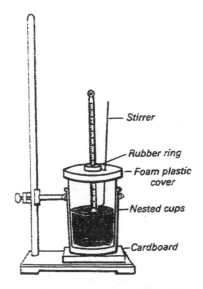

Figure 9.1   A calorimeter constructed from foam plastic cups and common laboratory equipment.

Pour the cold water into the calorimeter. After 5 minutes, read its temperature and the temperature of the warm water, estimating to within 0.1°C, applying any necessary correction for the No. 2 thermometer (2). Immediately transfer the warm water (the temperature of which has *just* been taken) to the calorimeter. Replace the lid and stir the mixture. Record the highest temperature reached (3).

Calculate the amount of heat the warm water loses (4) and the amount of heat the cold water gains (5), taking the density of water as 1.00 g/mL and the specific heat as 1.00 cal/g · °C. The difference is the heat that the calorimeter gains (6).

In all subsequent determinations made with this calorimeter, correct for the heat capacity of the calorimeter, which is the heat the calorimeter gains divided by the temperature rise; that is, cal/°C (7).

2. Neutralization of 0.50 M HCl and 0.5 M NaOH

Dry the calorimeter. Wipe the thermometers. Pipet 50.0 mL of 1.00 M NaOH into the calorimeter, using a safety pipet filler (see Part 3G). Replace cover and thermometer.

Pipet 50.0 mL of 1.00 M HCl into a dry beaker. When these solutions have reached nearly the same temperature (less than 0.5°C difference), record the readings (8). Pour the acid quickly into the calorimeter. Stir gently and record the maximum temperature reached (9).

Calculate the enthalpy of neutralization per mole of water formed (10). It can be assumed that the resulting NaCl solution has the same density and specific heat as water in this approximate determination. Write the ionic equation for the reaction (11), including the ΔH term.

3. Neutralization of 0.50 M NaOH and 0.50 M CH₃COOH

Repeat procedure in Section 2, using 1.00 M solutions of sodium hydroxide and acetic acid, after rinsing and drying the calorimeter and beaker and

pipets. Record the initial temperatures (12) and the maximum final temperature (13).

Calculate the enthalpy of neutralization per mole of sodium acetate formed (14). Assume that density and specific heat are the same for the sodium acetate solution as for water. Write the ionic equation, including the ΔH term (15).

4. Neutralization of 0.50 M NH₃ and 0.50 M HCl

Repeat procedure of Section 2 with 1.00 M solutions of ammonia and hydrochloric acid, after rinsing the vessels and pipets. Record the initial temperatures (16) and the final maximum temperature (17).

Calculate the enthalpy of neutralization per mole of ammonium chloride formed (18). Assume that density and specific heat are the same for ammonium chloride as for water. Write the ionic equation, including the ΔH term (19).

5. The Law of Hess (see Discussion)

The data obtained in this experiment are probably too imprecise to permit calculating a meaningful value of the enthalpy of neutralization of CH₃COOH (acetic acid) and NH₃ (ammonia). The principle, however, can be applied. Use the terms ΔH_1, ΔH_2, and ΔH_3 to represent the enthalpy of each reaction studied, and indicate the algebraic sign of each. Now write the ionic equations for the reactions in procedures 2, 3, and 4 in such a way that they add up to the equation for the reaction of CH₃COOH and NH₃ (20). Express the ΔH of this reaction algebraically in terms of ΔH_1, ΔH_2, and ΔH_3 (21). Convert your value, which is in kcal mol⁻¹ to kJ mol⁻¹. (1 cal = 4.184 J) (22)

(Optional.) You may wish to measure the enthalpy of reaction of CH₃COOH and NH₃(*aq*). If so, record the initial temperatures (23) and the final temperature (24) obtained.

Calculate the enthalpy of reaction per mole of ammonium acetate and water formed (25).

NAME _____ SECTION _____ DATE _____

1. Heat Capacity of the Calorimeter

(1) Correction required for thermometer No. 2 .... _____ °C

(2) Temperature of cold water .... _____ °C

     Temperature of warm water .... _____ °C

(3) Maximum temperature reached .... _____ °C

(4) Heat lost by warm water .... _____ cal

(5) Heat gained by cold water .... _____ cal

(6) Heat gained by calorimeter (4) − (5) .... _____ cal

(7) Heat capacity of calorimeter .... _____ cal/°C

2. Neutralization of 0.50 M HCl and 0.5 M NaOH

(8) Initial temperature of sodium hydroxide .... _____ °C

     Initial temperature of hydrochloric acid .... _____ °C

(9) Maximum temperature reached .... _____ °C

(10) Enthalpy of neutralization .... _____ kcal/mol

(11) Equation:

3. Neutralization of 0.50 M NaOH and 0.50 M CH₃COOH

(12) Initial temperature of sodium hydroxide .... _____ °C

     Initial temperature of acetic acid .... _____ °C

(13) Maximum temperature reached .... _____ °C

(14) Enthalpy of neutralization .... _____ kcal/mol

(15) Equation:

4. Neutralization of 0.50 M NH₃ and 0.50 M HCl

(16) Initial temperature of ammonia .... _____ °C

      Initial temperature of hydrochloric acid .... _____ °C

(17) Maximum temperature reached .... _____ °C

(18) Enthalpy of neutralization .... _____ kcal/mol

(19) Equation:

5. The Law of Hess

(20) Equations:

(21)

(22)

NAME ——————————————————————————— SECTION —————— DATE ——————————

(OPTIONAL)

(23) Initital temperature of acetic acid .... ——————— °C

   Initial temperature of ammonia .... ——————— °C

(24) Final temperature .... ——————— °C

(25) Enthalpy of reaction .... ——————— kcal/mol

NAME _____ SECTION _____ DATE _____

Answer the following questions before beginning the exercise:

1. Define the following terms:
   (a) enthalpy

   (b) ΔH (include units)

   (c) $\Delta H^{\circ}_{f298.15}$

   (d) exothermic reaction (include sign of ΔH)

   (e) endothermic reaction (include sign of ΔH)

   (f) specific heat

   (g) heat capacity

2. In the equation $Q = -C(T_2 - T_1)$, the *heat flow* Q is known as the change in enthalpy ΔH. This value, ΔH, is to be measured for each reaction carried out in this exercise.
   (a) If heat is *absorbed by a reaction*, carried out in the calorimeter, will the heat of the *calorimeter and contents* be lost or gained? (Underline correct answer.)

   (b) If heat is *evolved by a reaction*, carried out in the calorimeter, will the heat of the *calorimeter and contents* be lost or gained? (Underline correct answer.)

89

(c) The heat capacity, C, in the equation above of water (or dilute solutions) used in the calorimeter is the product of the mass and specific heat of water. What is the specific heat of water?

(d) Why must the heat capacity, C, be determined for the calorimeter itself?

3. Calculate Q (that is, ΔH), given the following data:

The neutralization of 50 mL of a certain 1 M acid by 50 mL of a certain 1 M base was carried out in a calorimeter. The initial temperature was 27°C, and the final temperature was 25°C.

Was this reaction exothermic or endothermic? Explain your answer.

10

Atomic Spectroscopy

OBJECTIVE

To measure wavelengths of the hydrogen emission spectrum and relate them to electron transitions; to identify some metals from the emission spectra of their ions.

DISCUSSION

When an electric discharge is passed through a monoatomic gas contained in a tube at low pressure, the atoms are excited; as they relax, the gas gives off light. The word *light* as we are using it refers to the *visible* portion of the electromagnetic spectrum (Fig. 10.1). (Atoms also emit radiation in other regions of the electromagnetic spectrum.) This light, when analyzed, shows a characteristic bright-line spectrum (its component colors). The instrument used for analysis is the *spectroscope;* it contains a prism, or grating, that separates the emitted light into its component wavelengths, to be observed and recorded (Fig. 10.2). A typical mounted spectroscope is shown in Fig. 10.3. The bright-line spectrum is characteristic of the atoms of the element making up the gas, and is called the *atomic spectrum* because the light is emitted from the excited atoms of the element. Each discrete line of the atomic spectrum corresponds to a single wavelength or a very narrow band of wavelengths of light emitted as the atoms return to lower energy levels in a discontinuous, quantized manner. This relation is expressed as follows:

$$E_{photon} = \Delta E_{electron} = \frac{hc}{\lambda} = h\upsilon$$

where E = energy
h = Planck's constant equal to 6.625×10^{-24} Joule $\cdot$ s
c = speed of light equal to 3.00×10^{10} cm/s
λ = wavelength in cm
1×10^{-8} cm = 1 angstrom unit (Å)
υ = frequency

After carefully measuring the relation of frequency to wavelength of the lines in

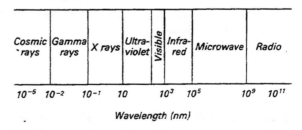

Figure 10.1 The electromagnetic spectrum. The scale is logarithmic.

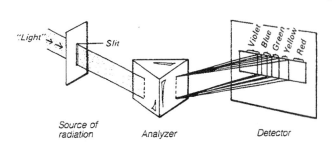

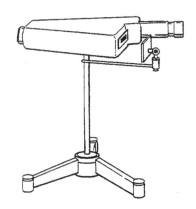

Figure 10.2 Essential components of a spectroscope. Not shown are the collimating lens interposed between the slit and the analyzer, and a lens between the analyzer and detector to focus the dispersed rays.

Figure 10.3 A typical mounted spectroscope.

the visible spectrum of hydrogen, J. R. Rydberg found in 1890 that the frequencies are related by this general equation:

$$v = R\left(\frac{1}{n_1^2} - \frac{1}{n_2^2}\right)$$

where R = Rydberg constant, of the value 3.289×10^{15} s^{-1}
n_1 and n_2 = integers (n_1 is smaller than n_2)

The Rydberg equation is empirical; that is, it is derived from observation rather than theory.

In 1913, Niels Bohr provided an explanation for the regularities Rydberg had observed. According to Bohr, the electron in the hydrogen atom can have only energies permitted by the following equation:

$$E = \frac{-21.79 \times 10^{-19}}{n^2} \text{ Joule}$$

The value -21.79×10^{-19} Joule is the energy required to remove the electron from the level closest to the nucleus completely out of the atom, and n is an integer corresponding to the energy level in which the electron is located.

Radiation is emitted when the electron "falls" from an outer energy level to one closer to the nucleus. The energy radiated equals the energy difference between these two levels and can be calculated from this equation:

$$E = 21.79 \times 10^{-19} \frac{\text{Joule}}{\text{atom}}\left(\frac{1}{n_1^2} - \frac{1}{n_2^2}\right)$$

The Rydberg equation and the Bohr equation can be brought into excellent agreement by multiplying the Rydberg equation by Planck's constant h.

The simple Bohr equation is derived from a system involving only one equation. If more than one electron is present, repulsive forces between the electrons complicate the calculations to the extent that predictions are virtually impossible. Modern physics and chemistry have given up the notion of well-defined electron orbits. Consequently, since the Bohr model is inadequate for multielectron systems, it is also inadequate for explaining chemical bonding. Nevertheless, the Bohr theory remains a milestone in scientific progress, inasmuch as Bohr was able to account satisfactorily, on the basis of his theory, for the bright lines in several atomic spectral series.

PROCEDURE

Your instructor will assign either Sections 1 and 2 or 1 and 3. In either case you will first complete a calibration curve for your spectroscope by plotting wavelengths (or scale units) observed against known values as outlined in Section 1. If Section 2 is assigned, you will observe and record the wavelengths for the hydrogen spectrum, and relate these data on hydrogen energy levels to electron transitions. If Section 3 is assigned, you will analyze qualitatively solutions containing metal ions. Each metal ion has a characteristic emission spectrum. You will identify the unknown by comparing its spectrum with known spectra. The excitation source for this work is a Bunsen burner flame.

1. Calibration of the Spectroscope

If the spectroscope you are to use has a scale of arbitrary units or if the accuracy of the markings is in doubt, it must be calibrated against a known standard—in this case helium. Since spectroscopes differ greatly, operating information is to be obtained from the instructor. **Caution: Do not touch the wiring.** Your instructor will place a helium discharge tube directly in front of the slit and activate the tube.

Seven lines should be visible through the eyepiece; they can be sharpened by narrowing the slit. The wavelengths of these lines, in nanometers (nm), are 668 (red), 588 (yellow), 502 (green), 492 (green), 471 (blue-green), 447 (blue-violet), and 403 (violet).

Make a table of these values and the wavelengths of the lines you see (*1*). If your instrument has arbitrary units, give these instead of wavelengths.

Plot wavelengths observed (or scale units) against the known values listed (*2*).

Draw a smooth curve through these points. This is a calibration curve for the instrument used.

2. The Hydrogen Spectrum

Your instructor will substitute a hydrogen discharge tube for the helium tube.

Record the wavelengths observed (*3*).

Determine the corrected wavelength values from the graph prepared in Section 1 (*4*).

From the corrected wavelengths, calculate the energies, in Joules per atom, corresponding to each line (*5*). Although your instrument may not show the lines at 434 nm and 410 nm, these should be included in your calculation.

Identify each line in the spectrum with an electron transition (*6*). Assign integers representing the two energy levels in the Bohr equation, n_1 being the level closer to the nucleus and n_2 the energy level farther away—using your calculated values for E.

Compare the wavelengths obtained with those calculated from the Rydberg empirical equation (*7*) and from the Bohr theoretical equation (*8*).

3. Spectra of Metal Ions

For this part, work in pairs. The excitation source is the Bunsen burner flame. One partner will introduce into the flame the solution containing the metal ion to be identified; at the same time, the other partner will record the spectral lines. If the solution is to be introduced into the flame by means of a platinum wire, preclean the wire by burning it in a flame until no residual color is imparted to the flame, alternately dipping it into 12 M HCl to facilitate the cleaning.

Caution: If a spray bottle is used, do not spray toward the spectroscope or toward your neighbor.

Open the slit of the spectroscope fairly wide. Set the burner in front of the slit but far away enough to avoid burning the spectroscope.

Obtain samples of known metal chloride solutions. Solutions of 0.1 M (or more concentrated) NaCl, KCl, $CaCl_2$, $SrCl_2$, or $BaCl_2$ are suitable.

Light the burner, introduce the solution into the flame, and record the wavelengths of the spectral lines (*9*). (*Note:* If a platinum wire is used instead of a spray bottle, the wire must be cleaned prior to using each different solution.)

Using your calibration curve to correct these values (*10*), compare them with known values (*11*) obtained from a reference book (handbook).

Your instructor may issue an unknown containing one or more metal ions for you to identify, which you would do by comparing the spectra (*12*). A blue glass (cobalt glass) is used to mask out sodium interference if K^+ is used in the unknown.

NAME _____ SECTION _____ DATE _____

1. Calibration of the Spectroscope

(1) Data for calibration of the spectroscope:

Color	λ, Known (in nm)	λ, Observed (in nm or scale units)
Red	668	
Yellow	588	
Green	502	
Green	492	
Blue-green	471	
Blue-violet	447	
Violet	403	

(2) Calibration curve. (Use above data. Plot your curve on the graph paper below.)

2. The Hydrogen Spectrum

(3)–(6)

λ, Observed	λ, Corrected	Energy (Joules/atom)	Electron transition

(7), (8) Comparison of λ: (Show sample calculations.)

λ, Observed (Corrected)	λ, Calculated (Rydberg equation)	λ, Calculated (Bohr equation)

NAME _____ SECTION _____ DATE _____

3. Spectra of Metal Ions

(9)–(11)

Metal	Color	λ, Observed	λ, Corrected	λ, Known

(12) Unknown:

Answer the following questions before beginning the exercise:

1. What is the difference between an *emission* spectrum and an *absorption* spectrum?

2. What do the lines in an emission spectrum correspond to?

3. Convert 668 nm to angstrom (Å) units. (1 Å $= 10^{-10}$m; 1 nm $= 10^{-9}$ m.)

4. Why does the helium spectrum have so many more lines than the hydrogen spectrum?

5. Would you expect the spectrum of the He^+ ion to be as complex as the spectrum of He? Explain.

6. The ion H_2^+ has one electron. Will the spectrum for H_2^+ resemble the spectra for any of H_2, H, He^+, and He?

7. Which of the following can be expected to give a spectrum most nearly like that of the H atom: Li, Li^+, or Li^{2+}?

11

Relative Chemical
Activity of Some Metals

OBJECTIVE

To become familiar with a method for determining the relative activities of metals as reducing agents.

DISCUSSION

The relative activities of many metals can be inferred from observing the rapidity with which they reduce the hydrogen ion and release hydrogen from acids. Since the rate of hydrogen evolution depends on both the strength of the acid and the activity of the metal used, one can compare extremely active metals, such as sodium and potassium, by their action on the extremely weak proton donor water. Somewhat less active metals require water at high temperatures or acids stronger than water for appreciable hydrogen evolution. Hydrochloric acid is a far stronger acid than water. This means that certain metals that react only slowly or not at all with water will displace hydrogen rapidly from hydrochloric acid. It is the hydrogen ion that oxidizes the metals, and its concentration is far greater in dilute acids than in pure water. For those metals that are incapable of releasing hydrogen from acids, comparison can be based on their reducing action on cations; that is, one can compare their chemical activities by observing their ability to displace other metals from aqueous solutions of their salts. For instance, silver is known to be a stronger reducing agent than gold because it liberates metallic gold from solutions of gold salts:

$$3Ag + Au^{3+} \longrightarrow 3Ag^+ + Au$$

The activity, or electromotive, series (see Fig. 11.1) is very useful since it indicates the possibility of a reaction of a given metal with water, acids, salts of other metals, and oxygen. The order in which the elements are placed in the series depends somewhat on the conditions under which the activity is observed. The series shown in Fig. 11.1 is applicable to reactions in water.

PROCEDURE

In this exercise you will rank various common metals in order of their relative activities by observing the rate at which they evolve hydrogen gas from cold, distilled water or from dilute hydrochloric acid. You will rank those that do not release hydrogen from water or aqueous acid according to their ability to displace other metals from aqueous solutions of their salts, the more active metal being the one tending to lose electrons. For example, Fe is more active than Cu, and thus can displace Cu from an aqueous solution of a copper(II) salt as shown by the following net ionic equation:

$$Fe + Cu^{2+} \longrightarrow Fe^{2+} + Cu$$

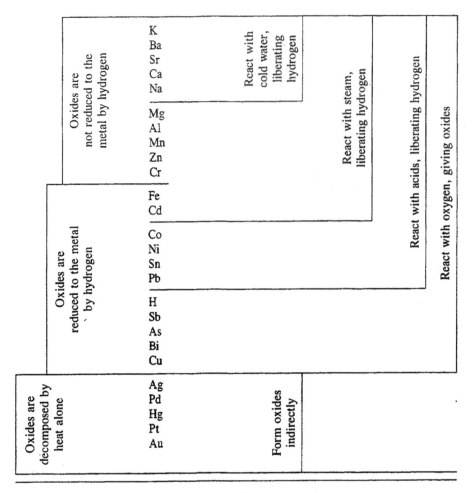

Figure 11.1 Activity, or electromotive, series of common metals.

1. Displacement of Hydrogen from Cold Water

Wrap a piece of sodium the size of a matchhead, obtained from your instructor, in a small bit of *dry* filter paper. **Caution: Use forceps when handling the sodium. Do not use your fingers because sodium reacts with water; your fingers may be moist.**

Fill a test tube with water and invert it in a beaker of water. Use a large beaker (600-mL) for ease in handling, with a minimum amount of water in it.

Hold the wrapped sodium with a forceps. Keep the test tube filled with water inverted in a beaker of water. Allow no bubbles of air to enter the test tube. Move the test tube more to the side of the beaker for access in handling. Now quickly insert the wrapped sodium under the mouth of the test tube without disturbing its position in the water.

After the reaction is complete, that is, when no more bubbles form, remove the paper from the tube using forceps, keeping the mouth of the tube under water.

Test the combustibility of the gas by carrying the inverted test tube to a flame. Is it combustible (*1*)? What is the gas (*2*)?

Test the solution in the beaker with red and blue litmus paper (*3*). Repeat this test with some sodium hydroxide solution from the reagent shelf and compare (*4*).

Drop a small piece of freshly cut calcium or lithium into a test tube containing 5 mL of water. Close the test tube with a *loose-fitting cork* until gaseous pressure develops, remove the cork, and immediately test the gas that has evolved with a flame (*5*). Test the solution in the test tube with litmus (*6*).

Place a piece of magnesium ribbon in a test tube containing 5 mL of distilled water. Is there any evidence of reaction (7)?

2. Displacement of Hydrogen from Acids

Set up six separate labeled test tubes in a rack. Place equivalent samples of the following metals in the test tubes: a 1 × 2-cm strip of copper, a piece of aluminum wire (that has been cleaned with sandpaper), a piece of tinfoil, a piece of iron wire (that has been cleaned with sandpaper), a piece of magnesium, and a piece of zinc.

Add to each test tube, *without delay*, 5 mL of dilute (6 M) hydrochloric acid. Observe the relative rates of evolution of hydrogen gas for a period of time for the different metals.

Using these observations, arrange these six metals in order of chemical activity, placing the most active first (8).

3. Displacement of Metals from Aqueous Solutions of Their Salts

A. Place a bright strip of copper in 5 mL of a 0.1 M solution of tin(II) chloride. Record your observation (9). Now place a bright piece of tinfoil in 5 mL of a 0.1 M solution of copper(II) sulfate. Record your observation (10). Which metal is more active, copper or tin (11)?

B. Add one drop of mercury from a dropper-bottle to 5 mL of a 0.1 M solution of silver nitrate and determine the relative chemical activity of the two metals, mercury and silver (12). **Caution: Return the used mercury to the container provided; do not discard in the sink.**

C. Now devise tests to determine the order of activity of the metals silver, mercury, tin, and copper. Carry out these tests and note the results (13).

SUMMARY

On the basis of your findings in this exercise, arrange all the metals tested in order of their activity (most active first) (14).

NAME _____ SECTION _____ DATE _____

1. Displacement of Hydrogen from Cold Water

(1)

(2)

(3)

(4)

(5)

(6)

(7)

2. Displacement of Hydrogen from Acids

(8) Order of chemical activity:

3. Displacement of Metals from Aqueous Solutions of Their Salts

(9)

(10)

(11)

(12)

(13) Description of tests and results:

Summary

(14)

THOUGHT QUESTIONS EXERCISE 11

NAME ______________________ SECTION __________ DATE __________

Answer the following questions before beginning the exercise:

1. Write ionic equations for the reactions, if any, of HCl with
   (a) Al

   (b) Sn

   (c) Fe

   (d) Mg

   (e) Zn

   (f) Cu

2. List the arrangement of chemical activity of these metals from a reference book.

3. Explain how displacement reactions might be used to determine the relative chemical activities of the nonmetallic elements bromine, chlorine, and iodine.

4. Having available CuO, Ag_2O, and any of the metals, outline a procedure by which you could demonstrate which is more active, copper or silver.

12

Chemical Bonding: General Concepts

OBJECTIVE

To illustrate ionic and covalent chemical bonding using Lewis structures; to demonstrate common structural features using molecular models; to illustrate the connection between structures and simple models for chemical bonds.

DISCUSSION

The electrons involved in bond formation between atoms are the *valence* electrons; that is, the electrons in the outermost (or in some transition elements, next to the outermost) energy level of the neutral atom. The number of valence electrons can be identified by the position of the element in the Periodic Table. Valence electrons are usually represented by dots written next to the symbol for the atom; for example, H· or Na·. These are called "electron-dot" formulas. In the case of H·, the single dot represents the single electron that the neutral hydrogen atom possesses; in the case of Na·, the single dot represents the electron that the neutral sodium atom, $^{23}_{11}$Na, has in its *outermost* energy level. The other electrons of the sodium atom are not designated because they are not involved in ordinary chemical reactions and bond formation.

When an atom that loses electrons easily, such as Na·, reacts with an atom that gains electrons easily, such as ·C̈l:, the reaction may be illustrated as follows: The single valence electron of sodium is transferred completely to chlorine, resulting in sodium ion, Na$^+$, and chloride ion, :C̈l:$^-$. A compound between these two ions is formed by electrostatic forces between the ions; namely, by an *ionic bond*.

$$\text{Na}^+ \ + \ :\ddot{\text{C}}\text{l}:^- \ \longrightarrow \ \text{Na}^+:\ddot{\text{C}}\text{l}:^-$$

| Sodium ion | Chloride ion | Sodium chloride, an ionic compound |

Electron transfer to form oppositely charged ions is favored when the atoms involved differ greatly in electronegativity. Typically, ionic bonds are found in compounds formed by the reaction between a metal in the Periodic Table of Groups IA and IIA (or transition metals), and a highly electronegative nonmetal of Groups VIA and VIIA. In such reactions there is a tendency for the atoms to acquire particularly stable configurations (an *octet* structure, ns^2np^6).

The nature of the bond formed when two atoms such as hydrogen combine to form the molecule H₂ may be represented as follows:

$$\text{H}\cdot \ + \ \text{H}\cdot \ \longrightarrow \ \text{H}\text{:}\text{H}$$

| Hydrogen atoms | Hydrogen molecule |

The pair of dots (electrons) in the hydrogen molecule indicates a *shared* pair of electrons, or a *covalent bond*. Covalent bonds have varying degrees of polarity, bond energy, and bond length—these properties vary with the nature of the atoms that share the electron pair.

Different theories and variations of theories have been proposed to explain covalent bonding. The stability of the covalent bond was set forth by G. N. Lewis, who proposed that atoms, by sharing electrons, can acquire a noble-gas configuration (often referred to as the *octet* rule; there are exceptions). Lewis structures can be written for most molecules (and ions), thereby exemplifying the number of shared electron pairs and unshared electron pairs. For example, the Lewis structure for ammonia, NH_3, is

$$H—\overset{..}{N}—H$$
$$|$$
$$H$$

This structure is arrived at in the following manner: Nitrogen has three bonds with hydrogen indicated by the single lines; nitrogen "brings in" five valence electrons; each hydrogen "brings in" one—a total of eight. Subtracting two electrons for each bond drawn in the skeleton structure leaves nitrogen with one unshared pair.

Lewis structures, together with the valence bond theory, and the concept of hybrid atomic orbitals, are useful in the prediction of the "geometry" (see text for details) of the molecule. Along with these ideas, the valence shell electron-pair repulsion (VSEPR) theory must be employed: The electron pairs surrounding an atom are oriented to be as far apart as possible; the electron pairs associated with the valence shell of a central atom either are bonding pairs or are unshared pairs that occupy a restricted region of space shaped rather like that occupied by the bonding pairs. Thus, for

$$H—\overset{..}{N}—H$$
$$|$$
$$H$$

with three bonds and one unshared pair of electrons; that is, a total of four regions of high electron density, the geometry (shape) would be trigonal pyramidal. The electron density of the unshared pair would not "appear" in the geometry of the molecule.

PROCEDURE

In this exercise you will write valence *symbols* by placing dots representing the correct number of valence electrons next to the symbol for the element. You will represent valence electron *formulas* (Lewis structures) for *ionic compounds* by combining the symbols for the particular positive and negative ions. A positive charge (+) indicates one or more electrons have been donated by an atom; a negative charge (−) indicates one or more electrons have been accepted by an atom. You will represent valence electron formulas (Lewis structures) for *molecular compounds* by combinations of valence electron symbols, with shared pairs of electrons designating covalent bonding.

You will construct some geometric models from ball-and-stick or plastic Minit model kits for a few simple molecules, based on Lewis structures plus the VSEPR theory. Remember, as far as molecular geometry is concerned, a double (two shared electron pairs) or a triple (three shared elec-

tron pairs) bond has the same representation as a single (one shared electron pair) bond.

1. Write valence electron *symbols* for the atoms of Period 3 of the Periodic Table: Na, Mg, Al, Si, P, S, Cl, and Ar (*1*).
2. Write valence electron *symbols* for the ions that would be expected to exist, corresponding to the atoms listed above (*2*).
3. Write valence electron *formulas* for sodium chloride and magnesium chloride, assuming ionic bonding (*3*).
4. Write a valence electron *formula* for silane, SiH_4, assuming covalent bonding (*4*).
5. Write a valence electron *formula* for the polyatomic ion, SO_4^{2-} (*5*).

Use Table 12.1 to direct you in answering 6 and 7.

6. First, draw Lewis structures for each of the following molecules or ions; then *make the model* representing its geometry. [If ball-and-stick models are used, the holes in the balls

(atoms) represent electron density (or valence) regions; the sticks represent bonds; the springs must be used for multiple bonds.] Record these data in the Report form, including sketches of the models and geometry name (6), on the basis of bonds.

H_2O, NH_3, NH_4^+, H_2, $H_2C=CH_2$, BF_3, BeF_2, CH_4, CO, N_2

7. Classify the geometry of the following species, on the basis of bonds plus unshared pairs of electrons. Refer to examples in Table 12.1 (7).

ICl_3, ClF_2, SF_6, ClF_5

8. What has to be the geometry of a diatomic molecule (8)? Explain.

TABLE 12.1 MOLECULAR STRUCTURES BASED ON THE VALENCE SHELL ELECTRON PAIR REPULSION THEORY

Regions of high electron density (Bonds and unshared pairs)	Molecular structures and examples (Chemical bonds are indicated in black; unshared pairs are gray)
Three: trigonal planar arrangement of bonds and/or pairs	3 Bonds / 0 Unshared pairs / Trigonal planar / CO_3^{2-}, BF_3, NO_3^- 　　　2 Bonds / 1 Unshared pair / Angular (120°) / NO_2^-, ClNO
Four: tetrahedral arrangement of bonds and/or unshared pairs	4 Bonds / 0 Unshared pairs / Tetrahedral / NH_4^+, CH_4　　3 Bonds / 1 Unshared pair / Trigonal pyramidal / H_3O^+, PCl_3, NH_3　　3 Bonds / 2 Unshared pairs / Angular (109.5°) / NH_2^-, H_2O
Five: trigonal bipyramidal arrangement of bonds and/or unshared pairs	5 Bonds / 0 Unshared pairs / Trigonal bipyramidal / PF_5, $SnCl_5^-$　　4 Bonds / 1 Unshared pair / Seesaw / SF_4, ClF_4^+　　3 Bonds / 2 Unshared pairs / T-shaped / ICl_3, ClF_3　　2 Bonds / 3 Unshared pairs / Linear / I_3^-, ClF_2^-
Six: octahedral arrangement of bonds and/or unshared pairs	6 Bonds / 0 Unshared pairs / Octahedral / PCl_6^-, SF_6, IF_6^+　　5 Bonds / 1 Unshared pair / Square pyramidal / IF_5, XeF_5^+　　4 Bonds / 2 Unshared pairs / Square planar / ICl_4^-, XeF_4

NAME _____ SECTION _____ DATE _____

(1) Valence electron symbols for the atoms of Period 3:

Na·   ·Mg·   ·Al: :Si: :P: :S: :Cl:

:Ar:

(2) Valence electron symbols for ions corresponding to (1):

Na^{+1} Mg^{+2} Al^{+3} Si^{+4} :Si:$^{-4}$:P:$^{-3}$:S:$^{-2}$

:Cl:$^{-1}$

(3) Valence electron formulas for

         Sodium chloride   *Magnesium chloride*

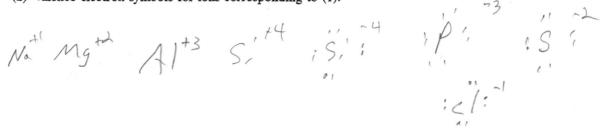

$MgCl_2$

Na—Cl: :Cl—Mg—Cl:

(4) Valence electron formula for SiH_4:

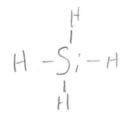

```
        H
        |
   H — Si — H
        |
        H
```

(5) Valence electron formula for SO_4^{2-}: *gives 2 more e⁻*

NAME _____ SECTION _____ DATE _____

(7)

| | Regions of high electron density | | | Molecular geometry |
	Unshared pairs	Bonds	Total	
ICl_3				
ClF_2				
SF_6				
ClF_5				

(8)

NAME _____ SECTION _____ DATE _____

Answer the following questions before beginning the exercise:

1. Define the term *valence:*

2. Draw an electron dot structure for sulfur:

$$\cdot \ddot{S} \vdots$$

3. Draw a Lewis structure for H_2SO_4:

4. Describe briefly what is meant by the VSEPR theory:

5. For the following questions, consult your textbook if necessary.
   (a) Write an electron configuration for the carbon atom:

   (b) Write an atomic orbital diagram for the *unhybridized* carbon atom:

   (c) Write an atomic orbital diagram showing the *hybridized* carbon atom:

13

Dimensions of a Molecule

OBJECTIVE

To become familiar with a method for determining molecular size from simple experimental data.

DISCUSSION

The type formula of a carboxylic acid is

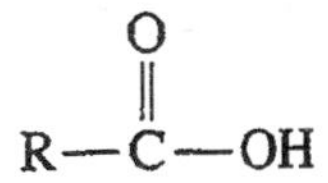

$$
\begin{array}{c}
\text{O} \\
\parallel \\
\text{R—C—OH}
\end{array}
$$

where R is any alkyl (hydrocarbon) group, and the remainder of the structure is the carboxylic acid group. The hydrocarbon portion of the molecule is insoluble in water; however, the carboxylic acid group forms hydrogen bonds with water molecules and is thus soluble. Hydrogen bonding results from intermolecular attraction between an available H atom on one molecule and an O, N, or F atom of an adjoining molecule. For example, hydrogen bonding in water may be exemplified in the following manner:

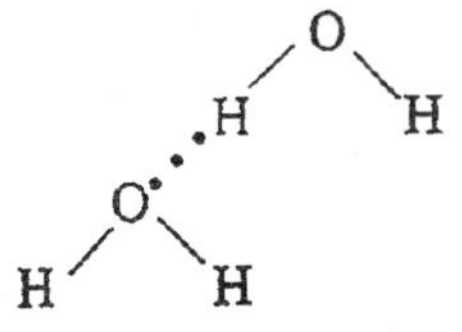

$$
\begin{array}{c}
\text{O} \\
\text{.H} \quad \text{H} \\
\text{O} \cdots \\
\text{H} \quad \text{H}
\end{array}
$$

The single lines indicate one pair of shared electrons; that is, a covalent bond. The triple-dotted line indicates hydrogen bonding, an *intermolecular force* (attractive force between molecules). As a result, a carboxylic acid in water will orient itself with the carboxylic acid groups hydrogen-bonded with water molecules on the surface and the hydrocarbon chain above the surface. If a monomolecular film (that is, a layer one molecule thick) of carboxylic acid on water is compressed laterally, the molecules become parallel and in contact, side by side. From Avogadro's number and the weight of the sample of a particular carboxylic acid, one can determine the number of molecules present in the water. Knowing the area of the film and the volume of the carboxylic acid, one can calculate the length of the molecule. If one views a molecule as a cylindrical rod, closely packed with other rods, it is easy to compute its diameter. Oleic acid, which is used in this exercise, has the formula $C_{17}H_{33}COOH$ and a density of 0.90 g/mL. The hydrocarbon portion of this carboxylic acid is the $C_{17}H_{33}$— group.

PROCEDURE

In this exercise you will use an indirect method of measuring the size of an oleic acid molecule, which obviously is too small to be measured directly. Follow the directions as outlined below and you will arrive at a value for the *length* of the oleic acid molecule, assuming that the oleic acid has spread out in a layer only one molecule in thickness, and that the thickness of the monolayer is equal to the length of the water-insoluble hydrocarbon portion of the molecule. A value for the *diameter* of the molecule is also attainable by calculation, assuming the molecule is rod shaped (cylindrical).

Carefully measure the diameter of a 15-cm watch glass to the nearest 0.10 cm (*1*).

Thoroughly clean the watch glass by scrubbing with scouring powder. Rinse and then clean the watch glass with hot detergent solution. Wash the watch glass under the tap for about 2 minutes and finally with distilled water. From this point on, handle the watch glass only by the edge to prevent contamination.

Obtain or prepare a pipet with a capillary tip as described in Part 3C. Determine the volume of the pipet and mark off the 1-mL and 2-mL divisions. A piece of gummed tape can be used to mark off the divisions.

Fill the pipet with ethanol; holding it vertically to make sure all the drops are of the same size, count the drops equivalent to 2 mL (*2*). There should be 100 to 150 drops per milliliter. *If there are fewer, empty the pipet and narrow the capillary tip in a flame.*

Calculate the volume per drop of ethanol (*3*).

Obtain about 5 mL of an oleic acid solution (0.3 mL of oleic acid in 100 mL of ethanol, accurately measured). (The volume per drop of the oleic acid solution is the same as for pure ethanol.)

Support the watch glass firmly in a horizontal position on an iron ring and add water until it is completely full. The watch glass should be level and free of motion.

Dust the surface with a very thin layer of lycopodium powder. If lycopodium powder is not available, fine chalk dust, talcum powder, or finely powdered sulfur may be substituted. **Caution: Lycopodium powder is flammable and should not be used around an open flame.**

Rinse the capillary pipet quickly with the oleic acid solution and fill it with the oleic acid solution. Support the pipet vertically just above the water. Allow just *one* drop to fall in the water. The ethanol evaporates, leaving oleic acid. Describe the effect on the lycopodium powder (*4*).

Add just *one* more drop and wait until the ethanol evaporates. Continue until the oleic acid film reaches the edge of the glass. An accurate measurement here is critical. Record the number of drops required (*5*).

Record the volume of oleic acid solution delivered, basing your calculation on the number of drops (*6*).

Next calculate the volume of *pure* oleic acid in the layer (*7*). The length of the oleic acid molecule is now determinable—it is assumed to be the thickness of the monolayer formed. Calculate the area of the monolayer [area $= \pi r^2$; r is available from Observation (*1*)]; calculate the thickness of the monolayer (equal to its volume divided by its area) (*8*). This value should equal the length of the oleic acid molecule.

How many oleic acid molecules are in the layer (*9*)? Assuming the molecules are rod-shaped and packed side by side as close as possible, what is the diameter of the molecule (*10*)?

REPORT EXERCISE 13

NAME ______________________ ______________________ SECTION __________ DATE __________

(1) Diameter of watch glass .... __________ cm

(2) Drops from pipet equivalent to 2 mL .... __________ drops

(3) Volume per drop of ethanol .... __________ mL

(4)

(5) Number of drops required .... __________ drops

(6) Volume of oleic acid delivered, based on the volume per drop (3) and the number of drops required (5) .. __________ mL

(7) Volume of *pure* oleic acid delivered, present in layer (A 0.3% solution of oleic acid was used.) .... __________ mL

(8) Length of oleic acid molecule .... __________ cm

(9) Number of oleic acid molecules in layer .... __________ molecules

(10) Diameter of oleic acid molecule (show calculation) .... __________ cm

NAME _____ SECTION _____ DATE _____

Answer the following questions before beginning the exercise:

1. (a) Show hydrogen bonding between three molecules of HF. (Use three dots
    to show hydrogen bonding and single lines to show covalent bonding.)

   (b) Show hydrogen bonding between a molecule of acetic acid, $CH_3C—OH$, with O double-bonded to C,
    and a molecule of water. [Use the same method as in 1(a).]

2. Why is it necessary to resort to an indirect method, such as this exercise, to
   show the dimensions of a molecule?

3. How many individual atoms are there in 11.5 g of Na? (Solution: Avogadro's
   number is involved. See your text as a reference.)

4. Show, by means of an equation, how volume divided by area can give you
   thickness (h):

14

Relation of Volume to Pressure at Constant Temperature: Boyle's Law

OBJECTIVE

To study the effect of pressure changes on gas volumes at constant temperature.

DISCUSSION

The volume of a given mass of a dry gas, at constant temperature, is reduced to half when the pressure applied to the gas is doubled. Conversely, the volume is doubled by a decrease in pressure to one-half. Robert Boyle, in 1660, summarized these observations in a form now known as Boyle's law.

This and other early discoveries categorized as *gas laws* can be readily explained in light of the kinetic-molecular theory of gases, a composite theory developed in the latter half of the nineteenth century. This theory, in a summarized form, states:

1. Gases are composed of molecules, the volume of which is very small compared with the total volume of the gas.
2. These molecules are in continuous motion and behave like elastic bodies on collision.
3. The average kinetic energy of the molecules of all gases increases with a rise in temperature. Average kinetic energy has the formula $\frac{1}{2}mu^2$, where m is the mass and u is the root mean square of the speed; this energy is the same for all molecules of all gases, regardless of mass, at the same temperature.

PROCEDURE

Because of the mercury involved, this exercise is suggested as a demonstration exercise, carefully monitored or carried out by the instructor. Two students at a time may be assigned to take readings, and alternating students may write the readings on the board. Each student in the class will copy down the data obtained, monitor the data being recorded, and submit an individual graph summarizing the results.

Obtain or assemble a Boyle's law apparatus similar to that shown in Fig. 14.1. If a gas-measuring tube as shown is not available, an inverted buret with a closed stopcock may be substituted. The amount of mercury required depends on the size of the measuring tube. **Before beginning the exercise, be certain the rubber tubes are tight, so that no mercury will be lost.** Also, throughout the exercise, take care not to upset the apparatus: **Spilled mercury is a hazard.** Should any mercury be spilled, collect all of it by drawing

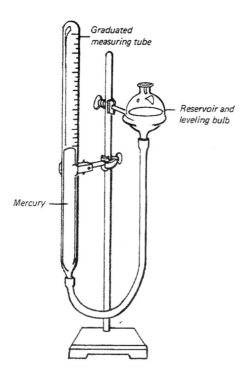

Graduated
measuring tube

Reservoir and
leveling bulb

Mercury

Figure 14.1 Determining the effect of pressure changes
on a volume of air at constant temperature.

it up with a medicine dropper, and place it in a
closed container.

The pressure on the gas trapped in the graduated
tube is increased by raising the leveling bulb and
decreased by lowering the leveling bulb. When the
levels of mercury in the bulb and the graduated
tube are the same, the pressure on the enclosed gas
is the same as the pressure of the atmosphere.
When the mercury level in the bulb is higher than
the level in the graduated tube, the pressure the
enclosed gas exerts equals the atmospheric pressure plus the pressure due to the mercury that lies
above the level in the graduated tube; that is, the
difference in mercury levels.

1. Adjust the height of the leveling bulb so that
the mercury stands at the same level in the
bulb and in the graduated tube. From the
graduated scale, determine the volume of the
enclosed gas as accurately as possible. Read

the laboratory barometer. Express the volume
of the gas in milliliters and the pressure in
millimeters.

2. Determine the volume of the enclosed gas for
six different positions of the leveling bulb—
three with the level of the mercury in the bulb
above the level of the mercury in the measuring tube, and three with the level in the bulb
below the level in the measuring tube. In each
case use a meter stick to measure the distance
between the mercury levels accurately in millimeters. When the level of mercury is lower in
the leveling bulb than in the measuring tube,
the pressure the gas exerts equals barometric
pressure minus difference in mercury levels. In
each case, determine the true pressure that the
trapped gas exerts.

3. Record your data for 1 and 2 in columns
headed with P and V, representing pressure
and volume (1). Multiply each value of P by
the corresponding value of V, and enter the
product in the column under K. Write an algebraic expression relating P, V, and the constant K (2). This is a mathematical expression
of Boyle's law. State the law in words (3).
Compute the average value of the constant K
(4) and precision obtained (5).

4. Make a plot of V (on the vertical axis) against
P (on the horizontal axis) on graph paper and
attach to the Report form. Choose a scale on
the graph paper that will use the maximum
space (6).

(*Optional.*) Students with access to an on-campus
computer science laboratory or a personal computer are encouraged to perform this plotting exercise using a spreadsheet, or similar computer program. Suitable programs are available through
several software development companies. Dis-
counts are often available to students purchasing
software for personal use. Contact your computer
science department, or local bookstore for more
information.

NAME _____ SECTION _____ DATE _____

(1)

EFFECT OF PRESSURE CHANGES ON GAS VOLUMES AT CONSTANT TEMPERATURE

	Differences in Hg levels, mm	Pressure P, corrected	Volume V, mL	K
1.	0			
2.				
3.				
4.				
5.				
6.				
7.				

(2) Algebraic expression relating P, V, and K (mathematical expression of Boyle's law):

(3) Statement of Boyle's law:

(4) Average value of K:

(5) Precision of value (4) (see Appendix for discussion of precision):

(6) Plot of V (vertical axis) against P (horizontal axis) (attach to report):

NAME _____ SECTION _____ DATE _____

Answer the following questions before beginning the exercise:

How does the kinetic-molecular theory account for the following facts?

1. The pressure of a gas doubles if its volume is reduced to half at constant temperature.

2. The pressure of a gas increases if it is heated in a closed, rigid container.

3. Gases deviate from the gas laws to the greatest extent at low temperatures and high pressures.

4. It is easier to liquefy some gases than others.

5. Hydrogen-filled balloons collapse more quickly than balloons filled with oxygen.

6. Nitrogen can be added to a rigid vessel already filled with oxygen.

7. The fraction of the total pressure that a gas in a mixture of gases exerts equals its fraction of the total number of moles and not its percentage by weight.

15

Relation of Volume to Temperature at Constant Pressure: Charles's Law

OBJECTIVE

To experimentally verify Charles's law; to graphically extrapolate the experimental volume-temperature data to absolute zero.

DISCUSSION

The volume of a given mass of gas is directly proportional to its temperature on the Kelvin scale when the pressure is held constant. The law is expressed mathematically as

$$\frac{T_1}{T_2} = \frac{V_1}{V_2}$$

This generalization, arrived at by Jacques Charles in 1787, is known as Charles's law.

Charles discovered through experimentation that the volume of a gas at constant pressure increased by $\frac{1}{273}$ of its value at 0°C for each degree Celsius rise in temperature, and decreased in the same manner as the temperature fell below 0°C. Theoretically, the volume of an ideal gas would continue to decrease until it became zero. Actually, before the temperature of −273°C, or *absolute zero* on the Kelvin scale is reached, all gases become liquids or solids.

The relationship between the Kelvin and Celsius temperature scales is

$$K = °C + 273$$

Kelvin temperatures, by convention, are reported without a degree sign.

PROCEDURE

Since an oil bath is needed for this exercise, the instructor may choose to obtain the data as a demonstration, and then make it available to the class for the balance of the exercise.

Obtain a small capillary tube prepared by your instructor for this exercise. The tube is sealed on one end. With a small metric ruler, measure the inside length, L_1 (*1*), and the inside diameter (*2*) of the tube to 0.5 mm. Calculate the initial volume in mL (cm^3) of air in the tube, V_1, from these measurements, by means of the equation $V = \pi r^2 h$ (*3*). ($\pi = 3.14$, $r = \frac{1}{2}$ diameter, $h =$ length.)

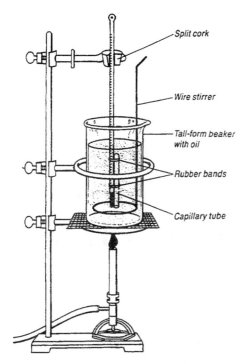

Figure 15.1 Determining the effect of temperature changes on a volume of air at constant pressure.

Attach the tube, open end down, to a thermometer by means of two rubber bands.

Set up a ring-stand assembly (similar to the one in Fig. 15.1) holding an oil bath; namely, mineral oil contained in a 200-mL tall-form Pyrex beaker. *The oil bath must be secured by a ring to prevent accidental spillage.* A split cork should be used in the clamp to hold the thermometer.

Lower the thermometer and tube into the oil bath until the tube is completely submerged. Include a wire stirrer. The thermometer must not touch the bottom of the beaker; the open end of the glass tube must be visible.

Gradually heat the oil bath until the bath temperature is approximately 70°C, occasionally stirring gently. Continue heating until a bubble of air is expelled from the tube. When this is seen to occur, *remove the burner* and read the *maximum* temperature (T_1) to 0.5°C (*4*). This may take several seconds.

Carefully raise the assembly so that most of the tube is out of the bath, but *keep the open end in the oil at all times.* Hold it in this position until the air column in the tube cools and contracts sufficiently to draw a column of oil about 4 to 5 mm long into the tube. Now, raise the thermometer and tube completely out of the bath and allow them to drain. Do not get oil on the table or floor. Allow the tube to cool to room temperature (T_2) (*5*). The inner meniscus of the oil column marks the end of the air column. Measure the length of the air column (L_2) (*6*). After the oil bath has cooled, return it to the designated place.

Calculate the final volume of air in the tube, V_2 (*7*). Calculate the value of T_1/T_2 (*8*) and V_1/V_2 (*9*). Are these values the same (*10*)? If not, calculate the percentage by which V_1/V_2 deviates from T_1/T_2 (*11*) as follows:

Percentage deviation

$$= \frac{(\text{difference in calculated ratios})}{T_1/T_2} \times 100$$

In this experiment, you compared the ratio of the temperatures with the ratio of the volumes of the expanded and contracted air column. Could you have based the comparison on the lengths of the air columns? That is, should $T_1/T_2 = L_1/L_2$? Give your answer and explain (*12*).

Construct a set of axes on graph paper with height of the air column on the vertical axis and temperature (°C) on the horizontal axis. Label the vertical axis from 0 to 10 cm and the horizontal axis from +100°C on the right to −300°C on the left. The two axes must intersect at 0°C and zero height.

Draw the best straight line through your experimental points. Extrapolate this line to the temperature axis. Record the temperature at which the extrapolated line intersects (*13*). Compare your value with absolute zero on the Kelvin scale (see Discussion) (*14*). Suggest a better means of obtaining this value (*15*). Attach your graph to the Report form (*16*). For an optional graphing method, see the suggestion found in Exercise 14, page 126.

NAME _____ SECTION _____ DATE _____

(1) Length (L_1) .... _____ mm

(2) Diameter .... _____ mm

(3) Initial volume (V_1) (show calculation) .... _____ mL

(4) Initial temperature (T_1) .... _____ °C _____ K

(5) Final temperature (T_2) .... _____ °C _____ K

(6) Length (L_2) .... _____ mm

(7) Final volume (V_2) (show calculation) .... _____ mL

(8) T_1/T_2, calculated .... _____

(9) V_1/V_2, calculated .... _____

(10)

(11) Percentage deviation (show calculation):

(12)

(13) Temperature .... _____ °C

(14)

(15)

(16) Graph (attach to Report)

NAME _____ SECTION _____ DATE _____

Answer the following questions before beginning the exercise:

1. State Charles's law in words.

2. Define the term "extrapolate."

3. Do the volumes of gases actually become zero at absolute zero? Explain.

4. Underline the correct answer: The relationship between temperature (K scale)
   and volume of gases is (direct, inverse).

5. (a) Convert 10°C to K:

   (b) Convert 303K to °C:

     139

16

Relation of Diffusion to Molecular Weight: Graham's Law

OBJECTIVE

To arrive at a relation between the rate of diffusion of a gas and its molecular weight (density).

DISCUSSION

Gases diffuse because the molecules of a gas are in rapid motion and the free space between molecules is very great. At the same temperature, molecules of different gases have the same kinetic energy (KE). Therefore, molecules of small mass move faster than those of large mass, and diffusion rates of gases are related inversely to their molecular weights and densities. This relationship may be shown as follows. For two different gases at the same temperature and pressure (where m is the mass of an individual molecule of a gas and u is its speed),

$$KE_1 = \tfrac{1}{2}m_1u_1{}^2$$

$$KE_2 = \tfrac{1}{2}m_2u_2{}^2$$

The kinetic energies of the two gases are equal. Hence,

$$\tfrac{1}{2}m_1u_1 = \tfrac{1}{2}m_2u_2{}^2$$

$$m_1u_1{}^2 = m_2u_2{}^2$$

Rearranging,

$$\frac{u_1{}^2}{u_2{}^2} = \frac{m_2}{m_1}$$

or

$$\frac{u_1}{u_2} = \sqrt{\frac{m_2}{m_1}}$$

We may assume that the rate at which a gas will diffuse R is proportional to the average speed of the gas molecules, and that the molecular weights M_1 and M_2 are proportional to their molecular mass. Thus,

$$\frac{R_1}{R_2} = \sqrt{\frac{M_2}{M_1}}$$

One mole of any gas at STP occupies 22.4 liters. Since the densities of any gas at STP can be calculated by dividing its gram-molecular weight by its volume (22.4 liters), the ratio of the densities is equal to the ratio of their molecular weights:

$$\frac{d_1}{d_2} = \frac{\dfrac{M_1}{22.4\,\ell}}{\dfrac{M_2}{22.4\,\ell}} = \frac{M_1}{M_2}$$

Therefore, Graham's law may be stated either in terms of the square roots of the molecular weights or the square roots of the densities of the gases.

$$\frac{R_1}{R_2} = \sqrt{\frac{M_2}{M_1}} = \sqrt{\frac{d_2}{d_1}}$$

PROCEDURE

In this exercise, you will work in pairs or in groups, as designated by your instructor. If, for some reason, the exercise must be repeated, use a different clean, dry tube if available, or clean and *thoroughly* dry the present one. A final drying with a hot air blast or cautious heating with a Bunsen flame is advisable. Read all of the directions before proceeding.

Clamp a clean, dry glass tube, 10 mm in diameter and 60 cm long, in a horizontal position. Have wads of cotton, two corks which fit the ends of the tube, and a stopwatch readily available.

One student adds 7 drops of concentrated hydrochloric acid (**Caution**) to one wad of cotton, and another student adds 7 drops of concentrated aqueous ammonia to another wad of cotton. **not inhale fumes.** Quickly, and simultaneously, that is, at the exact same time, the students place the two wads of treated cotton in opposite ends of the tube, and cork the ends of the tube. Start the stopwatch immediately. Stop the stopwatch when the white ammonium chloride smoke ring ap-

pears, indicating the meeting place of the two colorless gases. Record the time in seconds required for the gases to meet (*1*).

Measure the distance to 0.1 cm from the inner end of the cork on the ammonia end to the smoke ring deposit (*2*). Measure the distance to 0.1 cm from the inner end of the cork on the hydrogen chloride end to the smoke ring deposit (*3*).

Calculate the rate of diffusion of ammonia, $R_{NH_{3(g)}}$ (*4*), and of hydrogen chloride, $R_{HCl_{(g)}}$ (*5*), in cm/s.

Calculate the ratio $R_{NH_{3(g)}}/R_{HCl_{(g)}}$, using these experimental data (*6*).

Calculate the same ratio, using the molecular masses of the two gases (see Discussion) (*7*).

Calculate the percent of experimental error as follows:

% Error =

$$\frac{\text{difference in values of (6) and (7)}}{\text{value from Graham's law (7)}} \times 100 \;\; (8)$$

NAME _____ SECTION _____ DATE _____

(1) Time required for the gases to meet .... _____ s

(2) Distance of $NH_{3(g)}$ to meeting place .... _____ cm

(3) Distance of $HCl_{(g)}$ to meeting place ....;............. _____ cm

(4) $R_{NH_{3(g)}}$ .... _____ cm/s

(5) $R_{HCl_{(g)}}$ .... _____ cm/s

(6) $R_{NH_{3(g)}}/R_{HCl_{(g)}}$ from data .... _____

(7) $R_{NH_{3(g)}}/R_{HCl_{(g)}}$ from equation .... _____

(8) Percentage of error (show calculation):

NAME _____ SECTION _____ DATE _____

Answer the following questions before beginning the exercise:

1. What is meant by STP?

2. What is the molar volume of any gas at STP?

3. Write the equation for the formation of ammonium chloride (*s*) from hydrogen chloride (*g*) and ammonia (*g*).

4. How can you calculate the density of any known gas at STP?

5. What is meant by the "rate of diffusion" of a gas in terms of the kinetic molecular theory?

6. Calculate the ratio R_1/R_2 where hydrogen gas (H_2) is R_1 and helium (He) is R_2 (see Discussion). Which gas travels faster?

17

Solutions and Solubility

OBJECTIVE

To become familiar with some properties of solutions.

DISCUSSION

Solutions are defined as homogeneous mixtures. All solutions are characterized by homogeneity, absence of settling, and the molecular or ionic state of subdivision of the components. Because solutions differ in the physical state of both the solute (substance dissolved) and the solvent (dissolving medium), many kinds of solutions are possible.

The solubility of a gas in a given liquid is considered to be a specific property of the gas because a gas differs from other gases in the same liquid in solubility. *Miscibility* is a term used for liquids to indicate solubility of one liquid in another. Liquids that mix with water in all proportions are said to be completely *miscible* with water; either such liquids are ionic in solution or they are polar substances. *Immiscible* liquids in contact with each other form layers. The presence of a nonvolatile solute in a solvent (solid in a liquid) lowers the vapor pressure of the solvent, a so-called *colligative* effect, depending on the number of solute particles present in a given amount of solvent.

The dependence of solubility on temperature of any particular solid–liquid solution is apparent from a plot of its temperature–solubility curve.

PROCEDURE

Your instructor will assign the sections you are to perform. Determine this beforehand.

In this exercise you will observe some properties of three different kinds of solutions: gases in liquids, liquids in liquids, and solids in liquids. You may determine the solubility of an unknown solid in grams per 100 grams of water, for comparison with the handbook value. You may obtain data and construct a graph showing the solubility–temperature relationship for potassium nitrate in water, an example of a solid–liquid solution.

1. Gases in Liquids

Place some tap water in a beaker, and set up a ring-stand assembly for heating the water. Sus-

pend a thermometer in the water, by means of a clamp on the ring stand and a rubber band. The bulb of the thermometer should be completely covered by the water, but should not touch the bottom or sides of the beaker. Heat the water to about 50°C, and observe and record any changes you see in the water as it is heating (*1*). On the basis of these results, decide how the solubility of gases in liquids is affected by temperature changes, and record your answer (*2*).

2. Liquids in Liquids

Determine whether the following pairs of liquids are soluble (miscible) in each other by mixing 2-mL portions of them in a test tube; use a clean, dry test tube for each pair. Shake the mixture gently, and then allow to stand for a few minutes. Note and record whether layers form or whether complete miscibility exists: water and ethanol (*3*); water and glycerol (*4*); water and hexane (*5*); ethanol and glycerol (*6*); ethanol and hexane (*7*); glycerol and hexane (*8*). Account for your result in each case, by referring to the definitions of *miscibility* and *immiscibility* given in the Discussion (*9*).

3. Solids in Liquids

A. Effect of Solute on the Boiling Point of the Solvent.

Add 10 g of sodium chloride to 50 mL of water contained in a small beaker. Suspend the thermometer in the manner described in Section 1; heat until the solid is all dissolved and the solution is boiling freely. Record the boiling temperature (*10*). What effect has the solute on the boiling point and vapor pressure of the solvent (*11*)?

B. Determining the Solubility of an Unknown Solid.

Obtain 10 g of an unknown solid and pulverize it in a clean mortar. Add about 5 g of this solid to a flask, and shake vigorously with about 10 mL of water. If the solid dissolves completely under these conditions, add more solid in order to ensure a saturated solution.

Weigh a clean, dry evaporating dish to the nearest 0.01 g (*12*). Set up a filtration assembly (see Fig. 17.1). Do not wet the filter paper, but hold it in place. Filter the saturated solution. Weigh the evaporating dish with the solution to the nearest 0.01 g (*13*). Subtract (*12*) from (*13*) and record the weight of the solution (*14*).

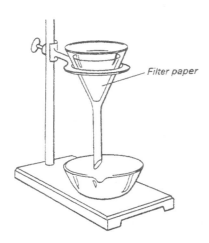

Filter paper

Figure 17.1   Filtration assembly.

Now place the weighed evaporating dish with the filtered solution on a wire gauze on a ring stand, and using a low flame, heat until the water (solvent) is evaporated, and the residue (solute) is nearly dry. Continue the evaporation over a steam bath, in order to prevent spattering and loss of solute, until dryness is complete. A hot plate may be used instead, lowering the temperature as the dryness nears completion.

Cool the evaporating dish with dried solute in it, and weigh to the nearest 0.01 g (*15*). Subtract (*12*) from (*15*) and record the weight of the solute (*16*).

Calculate the weight of the solvent, keeping in mind that a *solution* consists of two parts; namely, the *solvent* and the *solute* (*17*). Calculate, by ratio and proportion, the solubility of the solid in g per 100 g of water, using room temperature (*18*). What is the value listed for your unknown in the handbook (see instructor) (*19*)? Include the temperature listed.

C. Effect of Temperature on Solubility of a Solid.

Obtain a 200 × 25-mm test tube with a two-hole stopper equipped with a thermometer and a stiff wire stirrer bent into a loop at the bottom. The loop should be large enough to fit around the thermometer and yet not scrape against the sides of the test tube. The stopper should have a wedge removed from it so that the entire scale of the thermometer is visible.

Weigh 10 g of potassium nitrate to the nearest 0.1 g (*20*) and transfer it to the test tube. Clamp the test tube in place on a ring stand.

By means of a graduated pipet, add 5.0 mL of water to the test tube. Calculate the weight of the water, using 1.0 g/mL as the density of water (*21*). Heat the solution gently and not just in one spot, stirring constantly, until all of the potassium nitrate has dissolved. Continue stirring and allow the solution to cool. Note the approximate temperature at which crystals begin to form.

Heat the solution a few degrees above this temperature, allow to cool, and this time determine the temperature to the nearest 0.1°C at which *initial* crystallization occurs (*22*).

Add 5.0 mL more of water, heat to dissolve the potassium nitrate, and determine the temperature of crystallization as before (23). Repeat the determination after a third (24) and a fourth (25) addition of 5.0 mL of water.

The solubilities obtained are in terms of a fixed weight of KNO₃ and various volumes of water.

Recalculate in terms of g KNO₃ per 100 g of water (26). Tabulate the data and plot solubility versus temperature on graph paper; that is, plot temperature, °C, on the abscissa (horizontal axis) and solubility, g KNO₃/100 g H₂O, on the ordinate (vertical axis) (27). Read from the curve the solubility of potassium nitrate at 0°C and 100°C (28). Compare with the values listed in the handbook (29).

REPORT EXERCISE **17**

NAME ______________________ SECTION ________ DATE ____________

1. Gases in Liquids

(1)

(2)

2. Liquids in Liquids

(3)–(9)

Pairs of Liquids	*Miscible*	*Immiscible*	*Reason*
Water + ethanol			
Water + glycerol			
Water + hexane			
Ethanol + glycerol			
Ethanol + hexane			
Glycerol + hexane			

3. Solids in Liquids

A. Effect of solute on the boiling point of the solvent

(10)

(11)

B. Determining the solubility of an unknown solid

(12) Weight of evaporating dish .... ____________ g

(13) Weight of evaporating dish and solution .... ____________ g

(14) Weight of solution (13) − (12) .... _____ g

(15) Weight of evaporating dish and solute .... _____ g

(16) Weight of solute (15) − (12) .... _____ g

(17) Weight of solvent .... _____ g

(18) Solubility of unknown solid (#___) .... _____ g/100 g H_2O
   (Show calculations here.)   at ___ °C

(19) Solubility of unknown solid (handbook value) .... _____ g/100 g H_2O
                                                                          at ___ °C

C. EFFECT OF TEMPERATURE ON SOLUBILITY OF A SOLID

(20) Weight of KNO_3 .... _____ g

(21) Weight of 5.0 mL H_2O .... _____ g

(22)–(26) Data for solubility–temperature curve:

Trial	Volume of H_2O mL	Crystallization temperature, °C	Solubility, g KNO_3/100 g H_2O
1	5.0		
2	10.0		
3	15.0		
4	20.0		

(27) Plot of solubility–temperature curve (attach to Report).

(28) Solubility of KNO_3, from (22), at 0°C .... _____
                                        at 100°C .... _____

(29) Solubility of KNO_3, from handbook, at 0°C .... _____
                                        . at 100°C .... _____

NAME _____ SECTION _____ DATE _____

Answer the following questions before beginning the exercise:

1. Name the two components of any solution, generally speaking.

2. List three characteristics of solutions.

3. Give a *specific product* example of each of the three different kinds of solutions: gases in liquids, liquids in liquids, and solids in liquids.

4. In a liquid-in-liquid solution, what determines which liquid is the solute and which liquid is the solvent?

5. Define the terms "miscible" and "immiscible."

155

6. What effect does a solid, nonionizing solute have upon the freezing point of the solvent in which it was dissolved?

7. Experimentally, 2 g of a solid dissolves in 8 mL of water. Express this value in terms of the solubility of this solid in g per 100 g of H_2O.

Solutions of Electrolytes

OBJECTIVE

To become familiar with some of the phenomena underlying the theory of ionization and ionic dissociation.

DISCUSSION

Certain classes of substances, when dissolved in water, cause changes in the boiling and freezing points of water that yield anomalous values for molecular weight—the calculated molecular weight is smaller than the molecular weight corresponding to the simplest formula. Solutes of this type also show high electrical conductivity when dissolved in water. Arrhenius explained these facts brilliantly in terms of ionic dissociation, which remains, with modifications, the core of modern concepts.

The molal boiling-point elevation of water is 0.512°C; this means that 1 mole of *solute particles* dissolved in 1000 g of H_2O raises the boiling point of water by 0.512°C. Keep this information in mind while carrying out the experiment in Section 1.

PROCEDURE

In this exercise you will observe and compare properties of solutions of electrolytes (consisting of ions) and nonelectrolytes (consisting of molecules).

1. Effect of Solutes on the Boiling Temperature of Water

Determine the boiling temperature of a 4.8 M (28.4%) solution of acetamide as follows. Acetamide is an organic molecule with the following structure:

$$CH_3-C{\overset{O}{\diagup\!\!\!\backslash}}NH_2$$

Place 10 mL of the solution in a 200 × 25-mm test tube and suspend a thermometer so that its bulb is just above the liquid. Heat the solution to boiling and record its temperature within 0.2°C (*1*).

Determine and record the boiling temperature of a 4.8 M (28.1%) solution of sodium chloride in the same way (*2*).

Compare the change in boiling point due to sodium chloride with the change due to acetamide (*3*). Since the molarity of these solutions is the same, why do they differ in boiling point (*4*)?

2. Electrical Conductivity

Your instructor may elect to carry out this procedure as a demonstration.

A. Obtain a conductivity apparatus. Check with your instructor. One type is characterized in Fig. 18.1. **Caution: In using it, be careful not to touch the electrodes with your hand while the current is on. Otherwise, there is a danger of electrical shock. Between trials disconnect the plug from the ac outlet or throw the switch.**

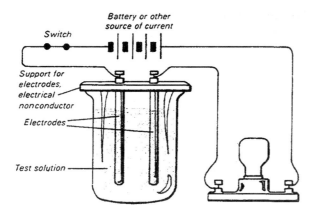

Figure 18.1 Apparatus for testing the electrical conductivity of solutions. A bright light indicates a good conductor.

A solid-state conductivity indicator powered by a 9-volt battery (see Fig. 18.2) may be available in your laboratory. This type of unit is safe. Varying intensities of the bulb mounted on the base indicate the conductivity of a liquid or solid.

A blinking LED (light-emitting diode) conductivity indicator, a safe unit requiring test solutions of small volume, may also be used. See your instructor for details of operation of this type.

Once you have obtained your unit and understand its use, rinse the electrodes with distilled water, and also the 50-mL beaker to be used. Place about 10 mL of each of the following aqueous solutions or liquids *in turn* in the beaker, then raise the beaker until both electrodes touch the surface of the liquid. Rinse the electrodes and beaker with distilled water before each trial.

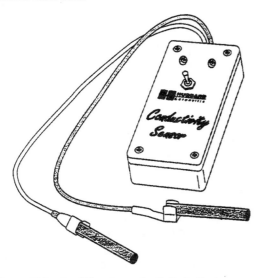

Figure 18.2 A solid-state conductivity indicator.

Classify each liquid as a good conductor, a poor conductor, or a nonconductor by checking the appropriate category *(5)*. The solutions are all 0.1 M:

1. Distilled water
2. Copper(II) sulfate
3. Hydrochloric acid
4. Sucrose (cane sugar)
5. Ethanol
6. Glycerol
7. Acetic acid
8. Sodium hydroxide
9. Magnesium nitrate
10. Sodium chloride

Your instructor may demonstrate the conductivity of a toluene solution of dry hydrogen chloride *(6)*. If so, then compare it with the conductivity observed for the water solution of hydrogen chloride in trial 3, and account for the result *(7)*.

B. Place about 2 g of dry, powdered potassium nitrate into a porcelain crucible supported by a triangle on a ring stand. With the conductivity apparatus disconnected, attach nichrome wires to each electrode and lead the wires into the crucible. **Caution: Do not allow the wires to touch each other.** Connect the apparatus and test the conductivity of the solid *(8)*.

Now heat the crucible with a Bunsen flame. Observe the lamp as the melting point is approached and passed; record your observation *(9)*.

Compare the result with the result from the unheated substance *(10)*. How would you explain the behavior of the solid *(11)*? What condition necessary for the conduction of current is indicated here *(12)*?

Wash the crucible, dry it, and place in it about 2 g of solid acetamide. **Caution: Do not allow the acetamide to contact the skin.** Test its conductivity as you heat the powder gently *(13)*. Does acetamide in the molten state conduct electricity *(14)*?

Compare this result with the result from the heated potassium nitrate *(15)*.*

Which of the substances tested seem to contain charged particles *(16)*?

---

*Dry sodium chloride and potassium nitrate are identical in behavior. Sodium chloride, however, melts at such an inconveniently high temperature that potassium nitrate is used here instead.

3. Conductivity Changes as a Result of Reaction

The apparatus should include an ammeter instead of a light bulb for better results, although estimations can be made by using bulbs of different wattages. Since mixing the reactants dilutes both, comparing conductivities of a single reactant and its mixture requires the concentrations being tested to be similar. This is done by diluting a portion of each reactant for measurement. Rinse the electrodes after each determination. **Use pipets equipped with safety fillers.**

A. Dilute 5.0 mL of 0.010 M HCl with an equal volume of water and record its conductivity (*17*). Do the same with 5.0 mL of 0.010 M NaOH (*18*). Mix 5.0 mL of 0.010 M HCl and 5.0 mL of 0.010 M NaOH and determine and record the conductivity of the mixture (*19*). Write the ionic equation for the reaction and account for the effect observed (*20*).

B. Follow the same procedure with 0.10 M H_2SO_4 and 0.10 M $Ba(OH)_2$. Record the initial conductivities (*21*) and the conductivity of the mixture obtained by adding 6 mL of the $Ba(OH)_2$ slowly to the acid (*22*). How does the conductivity change as the equivalence point is approached and passed (*23*)? Write the ionic equation for this reaction and account for the observed result (*24*).

C. Follow procedure **A** with 0.10 M solutions of acetic acid and ammonia. Record the separate conductivities of the acid and the base (*25*) and the mixture (*26*). Write the ionic equation for the reaction (*27*). Account for the difference in the results of procedure **A** and procedure **C** (*28*).

REPORT EXERCISE **18**

NAME ______________________ SECTION ________ DATE ________

1. Effect of Solutes on the Boiling Temperature of Water

(1) Acetamide .... ________ °C

(2) NaCl .... ________ °C

(3), (4)

2. Electrical Conductivity

A.

(5)

Liquid or solution (0.1 M)	*Good conductor*	*Poor conductor*	*Nonconductor*
1. Distilled water			
2. Copper(II) sulfate			
3. Hydrochloric acid			
4. Sucrose (cane sugar)			
5. Ethanol			
6. Glycerol			
7. Acetic acid			
8. Sodium hydroxide			
9. Magnesium nitrate			
10. Sodium chloride			

(6), (7) (Optional)

B.

(8) Conductivity of dry KNO_3 .... _____

(9) Conductivity of heated, dry KNO_3 .... _____
(10)–(12)

(13) Conductivity of solid acetamide .... _____
(14), (15)

(16) Substances tested that seem to contain charged particles:

3. Conductivity Changes as a Result of Reaction

A.

(17) Acid .... _____

(18) Base .... _____

(19) Mixture .... _____

(20) Ionic equation:

Explanation for result:

B.

(21) Acid .... _____

Base .... _____

(22) Mixture .... _____

(23)
(24) Ionic equation:

Explanation for result:

NAME _____ SECTION _____ DATE _____

C.

(25) Acid .... _____

Base .... _____

(26) Mixture .... _____

(27) Ionic equation:

(28) Reason for difference in parts A and C:

163

NAME _____ SECTION _____ DATE _____

Answer the following questions before beginning the exercise:

1. Summarize the classical Arrhenius theory of ionic dissociation.

2. Summarize the Debye–Hückel approach to the properties of solutions containing electrolytes.

3. Write equations for all the substances in Section (5) of the Exercise 18 Report that dissociate in solution. Balance and include correct charges on all ions.

165

19

Molecular Mass Determination by Solution Methods

OBJECTIVE

To gain experience in applying the quantitative laws of solutions to the determination of molecular mass.

DISCUSSION

A solute lowers the freezing point of a solvent, and provided that the solute is non-volatile, it raises the boiling point by an amount proportional only to the concentration of the solute particles. The chemical nature and masses of the solute particles are theoretically immaterial. These properties of solutions, which depend primarily upon the concentrations of solute particles rather than their nature, are termed *colligative* properties. The extent of change in freezing and boiling points that one mole of a molecular solute in 1000 g of solvent produces depends on the *solvent*. For the lowering of the freezing point in dilute solution, the relationship may be expressed as

$$\Delta T_f = k_f \times m$$

where ΔT_f is the freezing-point lowering in °C, m is the molality, and k_f is a constant for the particular solvent. For the raising of the boiling point in dilute solution, the relationship may be expressed as

$$\Delta T_b = k_b \times m$$

where ΔT_b is the boiling-point elevation in °C, m is the molality, and k_b is a constant for the particular solvent. For water, the molal freezing-point depression constant k_f is 1.86°C, and the molal boiling-point elevation constant k_b is 0.512°C. In applying this principle to determining molecular masses of solutes, it is often more convenient to use other solvents besides water: that is, solvents that have numerically greater constants and thus provide more easily measurable experimental differences. Biphenyl is a useful solvent for measurements at the freezing point, its freezing-point depression constant being 8.0°C. Heptane is a useful solvent for measurements at the boiling point, its boiling-point elevation constant being 3.43°C.

PROCEDURE

In this exercise you will perform either Section 1 or Section 2, as your instructor directs.

1. Molecular Mass by Freezing-Point Depression

Weigh a clean, dry 200 × 25-mm test tube to the

nearest 0.1 g (1). Fit the test tube with a two-hole stopper equipped with a thermometer and adjusted so that its bulb is about 2 cm from the bottom of the test tube. Through the other hole pass the stem of a stirrer made from stiff aluminum or nichrome wire bent into a loop at one end, the loop being large enough to easily accommodate the thermometer without contact, and yet small enough to move freely in the test tube.

Remove the stopper, and so on, and add approximately 15 g of biphenyl to the test tube; weigh the test tube and contents to the nearet 0.1 g (2). The difference between (2) and (1) is the weight of the biphenyl (3).

Now replace the stopper with its thermometer and stirrer, the circle of which surrounds the thermometer. Support an 800-mL beaker of water on a wire gauze and clamp the test tube in place (see Fig. 19.1), so that as much of it as possible is immersed in the water. Heat the water until the biphenyl is completely liquid. Shut off the burner, remove the water bath, and stir the melt rapidly as it cools. Record the temperature every 30 seconds, estimating to the nearest 0.1°C until the biphenyl is completely solid (4).

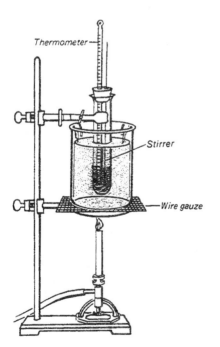

Figure 19.1 Apparatus for determining molecular mass by freezing-point depression.

Melt the biphenyl as before and repeat the determination to check your data (5). This value is the freezing point of biphenyl.

Weigh 3 g of naphthalene to the nearest 0.01 g (6). Melt the biphenyl in the water bath, and while the biphenyl is in the liquid state, transfer *all* the naphthalene to the test tube; stir until all the naphthalene has dissolved.

Remove the water bath and stir the solution as it cools. Record the temperature every 30 seconds until the mixture is completely solid (7).

Melt the mixture and repeat the determination as a check (8).

Plot the data obtained from the biphenyl alone and from the biphenyl plus naphthalene on one sheet of graph paper, and from the graphs decide on the freezing-point lowering due to the naphthalene (9).

Using the equation given in the Discussion, calculate the molecular mass of naphthalene; 8.0°C is the molal freezing-point constant for biphenyl (10).

The formula for naphthalene is $C_{10}H_8$. What is the actual molecular mass (11)? What is your percentage of error (12)?

Melt the mixture sufficiently to permit removal of the thermometer, and pour the contents into the provided receptacle; *do not pour the contents into the sink*. Exchange the test tube for another clean, dry test tube.

If an unknown is to be determined (see instructor), obtain about 3 mL or 3 g of the unknown and determine its molecular mass by the method used above for naphthalene. Record your data in the manner shown in the Report form.

Melt the mixture, withdraw the thermometer, and pour the contents into the provided receptacle. Turn in your test tube.

2. Molecular Mass by Boiling-Point Elevation

Weigh a clean, dry 200 × 25-mm test tube to the nearest 0.1 g (13). Add 20 mL of heptane to the test tube.

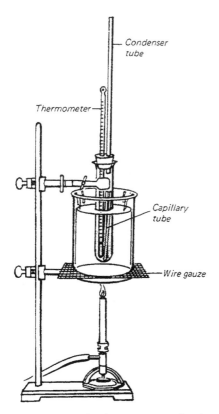

Figure 19.2   Apparatus for determining molecular mass by boiling-point elevation.

Now clamp the test tube so that it is half-immersed in an 800-mL beaker of water supported on a ring stand (see Fig. 19.2). Obtain or prepare several 15-cm capillary tubes sealed at one end, and place two in the liquid, open end down, to ensure even boiling. Fit the test tube with a two-hole stopper equipped with a 30-cm length of 8-mm glass tubing to serve as a condenser, and a thermometer adjusted so that its bulb is 1 cm from the bottom of the test tube. Heat the water bath. When a steady stream of bubbles issues from the bottom of the capillary tubes, estimate the thermometer to 0.1°C at 30-second intervals until a steady temperature is obtained (*14*). This is the boiling point of the heptane.

Remove the test tube from the water bath, take out the capillary tubes, and wipe the outside of the test tube. Weigh the test tube and contents (*15*); this can easily be done by standing the test tube and contents in a previously weighed small beaker and getting the weight by difference.

Obtain an unknown and dissolve about 4 g of it in the heptane and weigh again in the same manner (*16*).

Add new capillaries and determine the boiling point of this solution as before but exercise extreme care that only a small part of the heptane is in the condenser at any one time. Record the boiling point of the solution (*17*).

From the weight of the heptane (*18*), the weight of the unknown (*19*), and the boiling-point elevation (*20*), calculate the molecular mass of the unknown (*21*). Use the equation given in the Discussion.

NAME _____ SECTION _____ DATE _____

Complete the section assigned by your instructor.

1. Molecular Mass by Freezing-Point Depression

(1) Weight of test tube .... _____ g

(2) Weight of test tube + biphenyl .... _____ g

(3) Weight of biphenyl (2) − (1) .... _____ g

DETERMINATION OF FREEZING POINT OF BIPHENYL

(4) Temperatures (°C) at 30-s intervals   (5) Repeat of (4)

(6) Weight of naphthalene (or unknown) .... _____ g

DETERMINATION OF FREEZING POINT OF MIXTURE

(7) Temperatures (°C) at 30-s intervals   (8) Repeat of (7)

(9) Plot (attach to Report)

(10) Calculated molecular mass of naphthalene (or unknown):

(11) Actual molecular mass:

(12) Percentage of error:

2. Molecular Mass by Boiling-Point Elevation

(13) Weight of test tube .... _____ g

(14) Boiling point of heptane .... _____ °C

(15) Weight of test tube + heptane .... _____ g

(16) Weight of test tube + heptane + unknown .... _____ g

(17) Boiling point of mixture .... _____ °C

(18) Weight of heptane (15) − (13) .... _____ g

(19) Weight of unknown (16) − (15) .... _____ g

(20) Boiling-point elevation (17) − (14) .... _____ °C

(21) Calculated molecular mass of unknown:

NAME _____ SECTION _____ DATE _____

Answer the following questions before beginning the exercise:

1. Define the term "colligative properties."

2. What is the *molality* of a solution that contains 0.875 g of ammonia, NH_3, in 175 g of water?

3. Why is water not as useful a solvent as, for instance, heptane, in determining molecular mass by boiling-point elevation?

175

4. How much does the boiling point of water change when 1.00 g of sucrose, $C_{12}H_{22}O_{11}$, is dissolved in 50.0 g of water?

5. How much does the boiling point of water change when 1.00 g of salt, NaCl, is dissolved in 50.0 g of water?

6. (Underline correct answers)

   Solutes added to solvents (raise, lower) the freezing point of the solvent, and (raise, lower) the boiling point of the solvent.

20

Chemical Equilibrium

OBJECTIVE

To learn what factors affect the magnitude of equilibrium concentrations.

DISCUSSION

A principle of equilibrium, that of Le Châtelier, is that when a stress (for example, a change in pressure, in temperature, or in concentration) is applied to a system in equilibrium, the concentrations of the reactants will change in such a way that the stress is partially relieved.

Changes in pressure measurably affect systems in equilibrium only when gases are involved, and then only when the chemical reaction involves a change in the total number of gaseous molecules in the system. (According to Avogadro's law, equal volumes of gases at constant temperature and pressure contain equal numbers of molecules.) Consider the following hypothetical exothermic system:

$$A_2(g) + 2B_2(g) \rightleftharpoons 2AB_2(g) + \Delta H$$

If this equation represents not only the stoichiometric relation but the mechanism for the reaction, at equilibrium,

$$rate_{forward} = rate_{reverse}$$

and

$$k_{forward}[A_2][B_2]^2 = k_{reverse}[AB_2]^2$$

The "stress" of an increase in pressure on this hypothetical system would drive the reaction to the right as written; the chemical reaction that reduces the total number of molecules per unit of volume will be the one favored by an increase in pressure. The relation, directly stated, without resorting to Le Châtelier's principle, is that an increase in pressure on the system (because of a decrease in volume, say) increases the concentration of both reactants and products by the same factor. Since the forward reaction depends on the third power of the concentration, whereas the reverse reaction depends on only the second power, the rate of the forward reaction increases more than the rate of the reverse reaction. Thus there is a net reaction of A_2 and B_2 until the equilibrium is reestablished.

All chemical changes involve either the absorption of energy or the evolution of energy. Thus, in a system in equilibrium, an endothermic reaction (absorption of thermal energy) and an exothermic reaction (evolution of thermal energy) constitute the simultaneous forward and reverse reactions. When heat is the stress applied to a system in equilibrium, reaction in the endothermic direction is favored. Thus, when the temperature of a system is raised, the equilibrium is displaced so that thermal energy is absorbed; when the temperature is lowered, the equilibrium is displaced so that thermal energy is evolved. This generalization is known as van't Hoff's law, a special case of Le Châtelier's principle.

Consider, as an example of the effect of a change in concentration, the equilibrium

$$A + B \rightleftharpoons C + D$$

The symbols [A] and [B] represent molar concentrations of reactants to form molar concentrations of products [C] and [D] in the forward reaction. In the reverse reaction, [C] and [D] represent the molar concentrations of reactants, and [A] and [B] the molar concentrations of the products. Thus, for this system in equilibrium,

$$K_e = \frac{[C]_e[D]_e}{[A]_e[B]_e}$$

If, then, an additional quantity of the reacting molecule A is added to the system, the rate of the forward reaction will be temporarily greater than the rate of the reverse reaction as the concentrations of C and D increase. The rates of the forward and reverse reactions will become equal again; the ratio [C][D] to [A][B] will still equal the original K_e, although the equilibrium is said to have been shifted to the right. The system has shifted so that C and D are present in greater concentrations than originally, B is present in smaller concentration than originally, and A is present in larger concentration than before the stress of the addition of excess A.

PROCEDURE

In this exercise you will determine the effect of change in temperature on the equilibrium of a gaseous system, and the effect of change in concentration on the equilibrium of species in solution.

1. Effect of Change in Temperature

Obtain from the instructor a sealed tube of nitrogen dioxide gas and immerse it in ice water for several minutes (1). Account for the change in appearance (2).

Now warm the tube (3).

Write the equation for the equilibrium system, and, on the basis of your observations, indicate which reaction is exothermic (4). Show how the law of Le Châtelier can be applied to predict the effects of changes in temperature, total pressure, and partial pressures on this system (5).

2. Effect of Change in Concentration

A. To 50 mL of water, add one mL of 0.1 M iron(III) nitrate solution and 1.5 mL of 0.1 M ammonium thiocyanate (NH_4SCN) solution. Write the equation for the equilibrium established (6), and account for the color of the solution (7). Save this solution for the following tests.

To a portion of this solution, add a few drops of 0.1 M potassium thiocyanate solution and account for the change in the intensity of color (8). Name a substance that should favor the reverse reaction (9). Test your prediction (10).

B. Add 1 mL of 1 M $Fe(NO_3)_3$ and 1 mL of 1 M KSCN to 250 mL of water. If the solution is too dark, add more water. If too light, add 0.5 mL more of each of the original solutions. Write the equation for the equilibrium established (11). Save this solution for the following tests.

Place five clean, dry test tubes in a test-tube rack. Now place 10 mL of the solution you prepared in each of the five test tubes. Add to the test tubes the following solutions and account for the change in color, or other property.

1. 1 mL of 1 M $Fe(NO_3)_3$ (12)
2. 1 mL of 1 M KSCN (13)
3. 1 mL of 0.1 M $AgNO_3$ (14)
4. 3 drops of 0.1 M Na_2HPO_4 (15)
5. 2 mL of concentrated HCl (16) **Caution!**

Determine whether the enthalpy ΔH for the formation of the complex ion is positive or negative by placing one 10-mL portion of the original solution in ice and another in water at 50°C for 10 minutes (17).

NAME _____ SECTION _____ DATE _____

1. Effect of Change in Temperature

(1) Effect of cold:

(2) Account for change:

(3) Effect of warmth:

(4) Equation, including the ΔH term:

(5)

2. Effect of Change in Concentration

A.

(6) Equation:

(7) Account for color:

(8) Account for change:

(9)

(10)

B.

(11) Equation:

(12)–(16)

Solution added	Visible change in equilibrium (13)	Explanation for change
1		
2		
3		
4		
5		

(17) Effect of cold on equilibrium (11):

Effect of warmth on equilibrium (11):

Account for results:

NAME _____ SECTION _____ DATE _____

Answer the following questions before beginning the exercise:

1. State Le Châtelier's principle.

2. Give a brief summary of van't Hoff's law, a special case of Le Châtelier's principle.

3. Define the following terms:
   (a) *Endothermic reaction*

   (b) *Exothermic reaction*

4. What is the *molar concentration;* that is, $[AgNO_3]$, in a solution made by adding 1 g of $AgNO_3$ to 250 mL of H_2O?

181

21

Reaction Rate (Zinc–Hydrochloric Acid Reaction)

OBJECTIVE

To study the effect of changes in concentration on the rate of a chemical reaction.

DISCUSSION

The rate of a chemical reaction can be influenced markedly by changes in concentration of reactants, temperature, surface area, action of catalysts, and the nature of the reactant. This exercise is limited to the study of the effect of changes in concentration of reactants. At a fixed temperature and surface area, and in the absence of a catalyst (or in the presence of a fixed amount of a catalyst), the rate of a given reaction in solution is largely dependent upon the concentrations of the reactants. Collision between molecules and ions is necessary for the reaction to occur. With increased concentration of *any or all* of the reactants, the chances for collision are increased due to the greater number of molecules or ions per unit volume. Concentrations are expressed in moles per liter when the effect of concentration upon reaction rate is being considered.

Quantitative relationships between reaction rate and concentration for a one-step reaction may be expressed in the form of a rate equation. For a reaction of the type $A \rightarrow B + C$, the rate equation would appear as follows:

$$R = k[A]$$

where R is the reaction rate, k is the specific rate constant (which stays the same as long as the temperature of the system does not change), and the brackets mean *molar concentration of*. For a one-step bimolecular reaction, $A + B \rightarrow C + D$,

$$R = k[A][B]$$

with R and k being specific for this particular system. When more than one molecule of a reactant appears in the one-step reaction, for example, $2A \rightarrow B + C$, then

$$R = k[A]^2$$

the reaction rate being proportional to the square of the molar concentration of A. In general terms, the reaction rate is proportional to the product of the molar concentrations of each reactant, with the concentration of each reactant raised to the power equal to the number of molecules of that substance appearing in the equation.

PROCEDURE

In this exercise you will study the reaction between zinc and hydrochloric acid, and the effect on the reaction rate of increasing the concentration of hydrochloric acid, using a fixed amount of zinc—all other factors also being constant. Students should work in pairs or groups.

Set up the apparatus shown in Fig. 21.1. Fit the 250-mL Florence flask with a two-hole stopper equipped with a thermometer and a bent delivery tube. For glass-bending directions, see Part 3C, Figs. C.2 and C.3. The thermometer should reach to within 1 cm of the bottom of the flask.

Another tube, bent as shown, carries the evolved hydrogen to a gas-measuring tube (eudiometer). If a gas-measuring tube is not available, an inverted 100-mL graduated cylinder may be substituted. The rubber-tubing connection between these two glass tubes should be long enough to allow for flexibility, so that the stopper and its thermometer can be movable in order to add the zinc.

Fill the gas-measuring tube with water and invert it in a trough of water. Clamp the reaction flask and the gas-measuring tube to ring stands. All of the trials must begin with the acid at 35°C. **Caution: This reaction generates hydrogen gas, and should be done in a well-ventilated open area or in the hood.**

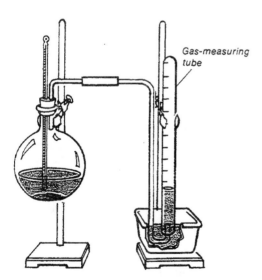

Figure 21.1 Assembly used in determining the effect of concentration on reaction rate of hydrochloric acid and zinc.

To ensure satisfactory results, it is essential to make your own samples of hydrochloric acid from the laboratory supply of 12 M concentrated hydrochloric acid. First make 3 M acid by mixing thoroughly 80 mL of 12 M hydrochloric acid with 240 mL of water. **Caution: Pour the acid slowly into the water while stirring.** Prepare 2 M acid by mixing 70 mL of 3 M acid with 35 mL of water. Prepare 1.5 M acid by mixing 50 mL of 3 M acid with 50 mL of water.

Weigh three 0.50-g samples of granulated zinc for use in the three trials below.

Place 100 mL of 3 M hydrochloric acid in the Florence flask and bring it to 35°C, with the stopper containing the thermometer *not closed* (rest it on the rim); otherwise air from the flask will enter the gas-measuring tube. **After the acid has been heated to this temperature, extinguish your Bunsen flame before proceeding.**

Obtain a stopwatch or one that can be read to seconds. Add the zinc to the acid all at once, being sure that all of it gets into the acid. Immediately close the stopper and note the time of addition. Allow the reaction to continue until exactly 25 mL of hydrogen have been collected and record the time required (*1*). Discontinue the reaction.

Rinse out the flask, fill the gas-measuring tube with water again, and repeat the determination, this time using 100 mL of 2 M hydrochloric acid at 35°C. Extinguish your flame before adding the zinc. Again, add the zinc, immediately close the stopper, and note the time of addition. Record the time required to collect 25 mL of hydrogen (*2*). Discontinue the reaction.

Rinse out the flask, fill the gas-measuring tube with water once again, and repeat the determination, this time using 100 mL of 1.5 M hydrochloric acid at 35°C. Extinguish your flame before adding the zinc. Add the zinc, immediately close the stopper, and note the time of addition. Record the time required to collect 25 mL of hydrogen (*3*). Discontinue the reaction.

Make a graph of concentration of acid (vertical axis) against time (horizontal axis) (*4*). What statement can be made about the effect of changes in the concentration of hydrochloric acid on the rate of reaction (*5*)?

NAME _____ SECTION _____ DATE _____

THE REACTION BETWEEN ZINC AND HYDROCHLORIC ACID

Trial	HCl (M)	Time in seconds required to collect 25 mL H_2
1	3	(1) _____
2	2	(2) _____
3	1.5	(3) _____

(4) Graph of [HCl] (vertical axis) against time (s) (horizontal axis) (attach to Report).

(5)

NAME _____ SECTION _____ DATE _____

Answer the following questions before beginning the exercise:

1. Read the Procedure for this exercise and decide what serious defect is inherent in the method used; briefly describe it.

2. Write the equation for the reaction studied in this exercise.

3. Assuming a one-step reaction, write the rate equation for this reaction (see Discussion).

4. Consulting your text, list three factors other than reagent concentration that influence reaction rate.

22

Reaction Rate (Iodine-Clock Reaction)

OBJECTIVE

To study the effect of changes in concentration on the rate of a chemical reaction.

DISCUSSION

This exercise is an alternative to Exercise 21, Reaction Rate (Zinc–Hydrochloric Acid Reaction). The Discussion in Exercise 21 is also applicable to this exercise.

PROCEDURE

In this exercise you will study the reaction between potassium iodate and sodium sulfite, with a starch solution added as an indicator, and the effect on the reaction rate of increasing the concentration of potassium iodate. The net reaction may be represented by the following net ionic equation:

$$2IO_3^- + 5SO_3^{2-} \longrightarrow I_2 + 5SO_4^{2-} + H_2O$$

Once the I_2 is formed, its presence is detected with the starch, forming a deep blue complex:

$$I_2 + starch \longrightarrow I_2 \cdot starch$$
$$\text{Blue complex}$$

Students should work in pairs or in groups. Each student or pair of students needs at least 150 mL of iodate solution, and 50 mL of starch-sulfite solution, allowing nothing for error or waste. The solutions must be prepared on the same day on which the exercise is performed. The instructor will either provide these solutions, or pre-assign their preparation as a student exercise.

Potassium iodate solution. Weigh out 4.3 g of potassium iodate and place it in a clean 1-L amber bottle. Half-fill the bottle with water and shake it until the salt is dissolved. Fill the bottle with water and mix the contents thoroughly.

Starch-sulfite solution. Dissolve 5 g of soluble starch in 100 mL of water by adding the starch to the cold water and bringing the mixture to a boil, stirring constantly. Cool the solution to room temperature, add 100 mL of water, and pour into a clean 1-L amber bottle. Dissolve 1.3 g of sodium sulfite in the solution; then add 13 mL of 3 M sulfuric acid and dilute to 1 L with water. Mix the solution thoroughly.

Use water at room temperature for the following work and make sure that the potassium iodate and starch-sulfite solutions are also at room temperature. Use two 10-mL pipets or 10-mL graduated cylinders for measuring the solutions, one for each solution; *do not get them mixed.* Measure the water carefully by graduated cylinder. Obtain a stopwatch or an ordinary watch that can be read to seconds. Have your Report form handy to record the data.

1. Measure 10 mL of the iodate solution and 80 mL of water into a 400-mL beaker placed on a sheet of white paper. Measure 10 mL of starch-sulfite solution into a clean test tube. With your watch ready, pour the contents of the test tube rapidly into the beaker, quickly mix the contents, and record the time (*1*). Look down into the beaker, which is on the

white paper, and be alert for the first appearance of the blue color. The blue $I_2 \cdot$ starch complex appears suddenly. Record to the nearest second the time when the blue color appears (*2*).

2. Repeat the experiment in a clean beaker using 20 mL of iodate solution, 70 mL of water, and 10 mL of starch-sulfite solution (*4*), (*5*).

3. Repeat the experiment in a clean beaker using 30 mL of iodate solution, 60 mL of water, and 10 mL of starch-sulfite solution (*7*), (*8*).

4. Repeat the experiment in a clean beaker using 40 mL of iodate solution, 50 mL of water, and 10 mL of starch-sulfite solution (*10*), (*11*).

Complete the tabulation of your data in the Report form. Make a graph of concentration of potassium iodate (vertical axis) against time required (horizontal axis), and attach to the Report form (*13*). What statement can be made about the effect of changes in iodate concentration on the rate of the reaction (*14*)? Summarize the results in an algebraic equation relating time required and concentration of iodate (*15*).

NAME _____ SECTION _____ DATE _____

The Iodine-Clock Reaction

Trial	KIO₃ (mL/100 mL)	Initial time	Final time	Time required for reaction (s)
1	10	(1)	(2)	(3)
2	20	(4)	(5)	(6)
3	30	(7)	(8)	(9)
4	40	(10)	(11)	(12)

(13) Attach graph of concentration of KIO_3 (vertical axis) against time (s) (horizontal axis).

(14)

(15) Equation:

NAME _____ _____ SECTION _____ DATE ___

Answer the following questions before beginning the exercise:

1. What is meant by a *net ionic* equation? (Consult your text, if necessary.)

2. The net ionic equation for this exercise is given in the Procedure.
   (a) Rewrite this net ionic equation as a *molecular* equation, showing all reactants and products as *compounds*.

   (b) Rewrite the same net ionic equation as a *total ionic* equation, showing *all cations and anions* written separately.

3. Consulting your text, list three factors other than reagent concentration that influence reaction rate.

23

Characterization of Compounds from Chemical Interaction (Acids, Bases, and Salts)

OBJECTIVE

To understand the chemical behavior of some combinations of cations and anions that can coexist in aqueous solution; that is, some of the solution chemistry of acids, bases, and salts.

DISCUSSION

Compounds that dissolve in water and increase the concentration of hydrogen ions, H^+, are *acids*. Hydrogen chloride, HCl, nitric acid, HNO_3, and sulfuric acid, H_2SO_4, are acids since they increase the hydrogen ion concentration in solution by the following reactions.

$$HCl(g) \xrightarrow{H_2O} H^+(aq) + Cl^-(aq)$$

$$HNO_3(l) \xrightarrow{H_2O} H^+(aq) + NO_3^-(aq)$$

$$H_2SO_4(l) \xrightarrow{H_2O} H^+(aq) + HSO_4^-(aq)$$

$$HSO_4^-(aq) \xrightarrow{H_2O} H^+(aq) + SO_4^{2-}(aq)$$

The terms (g), (l), (s), and (aq) refer to the gaseous, liquid, and solid states (phases), and to aqueous solutions, respectively.

As the equations are written, reference to the actual presence of the hydronium ion, H_3O^+, rather than the proton, H^+, is not necessary.

Compounds that dissolve in water and increase the concentration of hydroxide ions, OH^-, are *bases*. Potassium hydroxide, KOH, ammonia, NH_3, and sodium sulfide, Na_2S, are bases. Potassium hydroxide dissolves in water giving hydroxide ion; ammonia and sodium sulfide react with water to give hydroxide as indicated in the following reactions:

$$KOH \xrightarrow{H_2O} K^+(aq) + OH^-(aq)$$

$$NH_3 + H_2O \longrightarrow NH_4^+(aq) + OH^-(aq)$$

$$Na_2S + H_2O \longrightarrow 2Na^+(aq) + OH^-(aq) + HS^-(aq)$$

---

* This exercise was written by W. R. Robinson and G. M. Bodner, and is used at Purdue University.

A *salt* is an ionic compound composed of positive ions (cations) and negative ions (anions). The positive ions may be monoatomic cations other than H^+ such as K^+ or Ba^{2+} or polyatomic cations such as NH_4^+. The negative ions may be monoatomic anions other than OH^- such as Cl^- or S^{2-} or polyatomic anions such as NO_3^- or SO_4^{2-}. A salt that is soluble dissolves to give a solution containing the cations and anions of the salt.

Many acids, bases, and salts are soluble in water to give partially or completely ionized solutions. Thus the aqueous solution chemistry of the acids HCl, HNO_3, and H_2SO_4 corresponds to

$$H^+(aq) \quad \text{and} \quad Cl^-(aq)$$
$$H^+(aq) \quad \text{and} \quad NO_3^-(aq)$$
$$H^+(aq) \quad \text{and} \quad HSO_4^-(aq) \text{ plus } SO_4^{2-}(aq)$$

The common feature of the chemistry of acids in aqueous (water) solution is the presence of the hydrogen ion, $H^+(aq)$, in solution.

The solution chemistry of the bases KOH, $Ba(OH)_2$, and Na_2S corresponds to

$$K^+(aq) \quad \text{and} \quad OH^-(aq)$$
$$Ba^{2+}(aq) \quad \text{and} \quad 2OH^-(aq)$$
$$2Na^+(aq), \quad HS^-(aq), \quad \text{and} \quad OH^-(aq)$$

The common feature of the chemistry of aqueous solutions of bases is the presence of the hydroxide ion, OH^-, in solution.

In the case of salts, the cations are generally simple aquated metal ions such as Na^+, Zn^{2+}, Fe^{3+}, and so on, (also NH_4^+), while the anions may be simple, as in F^-, Cl^-, Br^-, I^-, and S^{2-}, or more complex, as in SO_4^{2-}, CrO_4^{2-}, $Cr_2O_3^{2-}$, ClO_4^-. The chemistry of the salts NaI, $Cu(NO_3)_2$, $BaCl_2$, and Na_2SO_4 corresponds to solutions of the following ions, respectively:

$$Na^+(aq) \quad \text{and} \quad I^-(aq)$$
$$Cu^{2+}(aq) \quad \text{and} \quad 2NO_3^-(aq)$$
$$Ba^{2+}(aq) \quad \text{and} \quad 2Cl^-(aq)$$
$$2Na^+(aq) \quad \text{and} \quad SO_4^{2-}(aq)$$

For aqueous solutions of completely ionized acids, bases, or salts, the reactions of the cations and anions are quite independent and they may be separately studied, correlated, and catalogued. For example, any solution containing $Ag^+(aq)$ will react with any solution containing $Cl^-(aq)$ to give solid, insoluble AgCl irrespective of what other cations or anions may be present. Thus a solution of $AgNO_3$ will react with a solution of HCl, NaCl, or $FeCl_3$, forming AgCl(s) in each case.

$$Ag^+(aq) + NO_3^-(aq) + H^+(aq) + Cl^-(aq) \longrightarrow$$
$$AgCl(s) + H^+(aq) + NO_3^-(aq)$$

$$Ag^+(aq) + NO_3^-(aq) + Na^+(aq) + Cl^-(aq) \longrightarrow$$
$$AgCl(s) + Na^+(aq) + NO_3^-(aq)$$

$$3Ag^+(aq) + 3NO_3^-(aq) + Fe^{3+}(aq) + 3Cl^-(aq) \longrightarrow$$
$$AgCl(s) + Na^+(aq) + NO_3^-(aq)$$

The net ionic reaction in each reaction is the following:

$$Ag^+(aq) + Cl^-(aq) \longrightarrow AgCl(s)$$

The solid formed when two solutions react is called a *precipitate*. Silver chloride, AgCl, is the precipitate in each of the reactions given above.

Many combinations of cations and anions can coexist in aqueous solution. Depending on the cation and anion present, the solution may be that of an acid, a base, or a salt. This experiment is designed to introduce some of the chemical behavior of these solutions and thus some of the solution chemistry of acids, bases, and salts.

PROCEDURE

In this exercise you will be provided with five test tubes, each containing a solution of an unknown compound. The tubes will be coded, but unlabeled. One test tube will contain a solution of an acid, either HCl, HNO_3, or H_2SO_4. One test tube will contain a solution of a base, either NH_3, KOH, or Na_2S. The third test tube will have a salt containing chloride, either NaCl, $BaCl_2$, or $CuCl_2$. The fourth test tube will have a salt containing nitrate, either $AgNO_3$, $Cu(NO_3)_2$, or $Fe(NO_3)_3$. The final test tube will contain a sodium salt, either NaI, Na_2SO_4, or Na_2CO_3.

Carefully read and follow the general directions for this exercise given in the Procedure. Your initial task will be to distinguish the acid and the base from the salts using the information found in the section on Observations, Tests, and Other Data that follows the Procedure. Then identify the particular acid, the particular base, and each salt in your unknowns. *Identify each unknown by the code number on the test tube.* In the Report form, give the code number (*1*), the identity of the unknown, a description in words of what you observe to take place as you characterize your unknown (*2*), and balanced equations for the reactions used to identify the unknown (*3*). It will not be necessary to do all the tests with an unknown in order to identify it conclusively (*4*). However, the exact number of tests necessary to determine your unknowns depends not only on what unknowns you have, but also on your individual manner of approach to the problem.

During this entire experiment it is imperative that test tubes and medicine droppers be scrupulously cleaned after each individual test. *Avoid contamination of the test solutions.*

Reagents and apparatus for the following tests will be available:

1. Determination of odor and color
2. Determination of acidity or basicity
3. Reaction with KOH
4. Reaction with $AgNO_3$
5. Reaction with $BaCl_2$
6. Reaction with NaCl
7. Reaction with $NH_3(aq)$
8. Reaction with Na_2S
9. Reaction with H_2SO_4

Obtain a sufficient number of small test tubes. Carry out *each* of the individual tests on your unknowns on a semimicro basis in the following manner: Using your *clean* medicine dropper, place a *5-drop sample* of your unknown (be certain to

record its code number) into a clean test tube which has had a final rinse with deionized (or distilled) water. Now, *slowly* add one of the test solutions, one drop at a time, with a second clean medicine dropper provided for this particular test solution, until a total of 10 drops of test solution has been added. The test solution is the solution you have previously decided upon in your analysis scheme. Describe briefly but accurately what you observe and record in the Report form. Using a fresh sample of your original unknown, record its behavior in the same way with whatever additional test solutions are necessary to confirm its identity. Continue in a similar manner with all of your unknowns.

OBSERVATIONS, TESTS, AND OTHER DATA

Observation of Odor and Color

Many useful observations can be made with the eye and nose.

Solutions of NH_3 and Na_2S have characteristic odors. If you are not familiar with these, *carefully* check the odors of the test solutions by gently sniffing the *closed* bottle or by fanning a little of the vapor above an open bottle toward your nose.

Solutions of certain metals containing salts have characteristic colors. Solutions containing Ni^{2+} (*aq*) are green; $Cu^{2+}(aq)$, blue; and $Fe^{2+}(aq)$, pale yellow. In addition, the precipitates that result when two solutions are mixed may have characteristic colors. These are indicated in the table of common insoluble salts.

Acidity-Basicity

Whether a solution is acidic or basic can be determined with the pH paper. The color of the paper when moistened with an unknown identifies the pH of the unknown. Remember that not only the acids and bases themselves, but also certain salts give acidic and basic reactions. Among the unknowns we can distinguish four classes of compounds that give acidic or basic solutions:

Strong Acids: pH < 3. HCl, HNO_3, H_2SO_4, and, for example, $HClO_4$ and H_3PO_4. These compounds give solutions containing $H^+(aq)$ due to the following reaction, which proceeds almost completely to the right:

$$HX \xrightarrow{\ H_2O\ } H^+(aq) + X^-(aq)$$

Acid Salts: $1 <$ pH < 7. $Fe(NO_3)_3$ and, for example, $FeCl_3$, $ZnSO_4$, and $Al_2(SO_4)_3$. Some of the metal ions in these solutions react with water to release a small amount of hydrogen ion according to the following reaction:

$$Fe^{3+}(aq) + H_2O \longrightarrow FeOH^{2+}(aq) + H^+(aq)$$

Basic Salts: $7 <$ pH < 11. Na_2CO_3, Na_2S, and, for example, Na_3PO_4 and Na_2SO_3. These compounds give solutions containing OH^- due to hydrolysis of the anion. Na_2S gives so much OH^- that we will consider it as a strong base in this experiment.

$$Na_2CO_3 + H_2O \rightleftharpoons \\ 2Na^+(aq) + OH^-(aq) + HCO_3^-(aq)$$

$$Na_3PO_4 + 2H_2O \rightleftharpoons \\ 3Na^+(aq) + 2OH^-(aq) + H_2PO_4^-(aq)$$

$$Na_2S + H_2O \rightleftharpoons \\ 2Na^+(aq) + OH^-(aq) + SH^-(aq)$$

Bases: pH > 11. $NH_3(aq)$, KOH, and, for example, $NaOH$ and $Ca(OH)_2$. These compounds give solutions containing OH^- due to reactions of the following sort:

$$NaOH(s) \xrightarrow{H_2O} Na^+(aq) + OH^-(aq)$$

$$NH_3(aq) + H_2O \rightleftharpoons NH_4^+(aq) + OH^-(aq)$$

Note that pH $= -\log[H^+(aq)]$, where $[H^+(aq)]$ indicates the hydrogen ion concentration. Thus, low pH values (1–3) indicate relatively high concentrations and thus indicate strongly acidic solutions. Since $[H^+(aq)][OH^-] = 10^{-14}$, low $H^+(aq)$ concentrations (high pH values, 11–14) indicate high OH^- concentrations or strongly basic solutions.

Solubility

The formation of a precipitate when two solutions are mixed indicates that a compound that is not soluble has formed. Observation of the formation of a precipitate coupled with the following guidelines about solubility can help identify unknowns.

1. All nitrates are soluble in water.
2. All common sodium, potassium, and ammonium salts are soluble in water.
3. All common halides (salts containing Cl^-, Br^-, or I^-) except those of Ag^+, Hg_2^{2+}, and Pb^{2+} are soluble in water.
4. All common sulfates (salts containing SO_4^{2-}) except those of Ba^{2+} and Pb^{2+} are soluble in water. Silver sulfate and calcium sulfate are slightly soluble, but at the 0.1 M concentrations used in this experiment *probably* will be soluble.
5. The sulfides (salts containing S^{2-}) of Na^+, K^+, NH_4^+, Ba^{2+}, and Ca^{2+} are soluble.
6. Ag^+, Cu^{2+}, Hg^{2+}, Ni^{2+}, and Zn^{2+} form soluble ammine complexes with excess ammonium hydroxide.

$$Ag^+(aq) + 2NH_3(aq) \longrightarrow Ag(NH_3)_2^+(aq)$$

$$Cu^{2+}(aq) + 4NH_3(aq) \longrightarrow \underset{\text{Deep blue}}{Cu(NH_3)_4^{2+}(aq)}$$

Evolution of Gases

Carbon dioxide can be seen bubbling out of a carbonate (CO_3^{2-}) solution when an acid is added.

H_2S is evolved when an acid is added to a sulfide solution, but bubbling may or may not be observed. However, this gas is easily detected by its distinctive odor.

(*Note:* A white suspension of colloidal sulfur is sometimes obtained when H_2SO_4 is added to Na_2S.)

COMMON INSOLUBLE SALTS

This table lists the color and appearance of the precipitates that form when a solution of a compound containing the indicated negative ion is mixed with a solution of a compound containing the indicated positive ion. For example, a solution of sodium bromide [$Na^+(aq) + Br^-(aq)$] reacts with a solution of silver nitrate [$Ag^+(aq) + NO_3^-(aq)$] to form a very pale yellow, heavy, curdy solid (precipitate) of AgBr, leaving a solution of $Na^+(aq)$ and $NO_3^-(aq)$ [sodium nitrate].

Bromides; Br^-
Ag^+	Very pale yellow; heavy; curdy
Hg^{2-}	(Mercuric)—white; soluble in excess Br^-
Pb^{2+}	White

Carbonates; CO_3^{2-}
Ag^+	White
Ba^{2+}	White
Ca^{2+}	White
Cu^{2+}	Light blue
Fe^{3+}	(Ferric)—light orange-brown; gelatinous ($Fe_2O_3 \cdot xH_2O$)
Hg^{2+}	(Mercuric)—dark red-orange, turning to brown
Ni^{2+}	Pale green; gelatinous
Pb^{2+}	White
Zn^{2+}	White; gelatinous

Chlorides; Cl^-
Ag^+	White; heavy precipitate
Pb^{2-}	White; moderately soluble; soluble in hot solution

Chromates; CrO_4^{2-}
Ag^+	Dark orange-red
Ba^{2+}	Light yellow
Cu^{2+}	Orange-brown
Hg^{2+}	Orange
Pb^{2+}	Bright yellow
Zn^{2+}	Bright yellow

Hydroxides; OH^-
Ag^+	With NaOH, gray-brown suspension. With NH_3, gray-brown; dissolves in excess NH_3 to colorless solution
Cu^{2+}	With NaOH, pale blue. With NH_3, pale blue; dissolves in excess NH_3 to deep blue solution
Fe^{3+}	Rust-colored
Hg^{2+}	(Mercuric)—with NaOH, yellow; only after excess base added. With NH_3, white; forms immediately; dissolves in excess NH_3
Ni^{2+}	With NaOH, pale green; gelatinous. With NH_3, pale green; gelatinous; dissolves in excess NH_3 to blue solution

Pb^{2+}	White
Zn^{2+}	With NaOH, white; gelatinous; With NH_3, white; gelatinous; dissolves in excess NH_3 to colorless solution

Iodides; I^-
Ag^+	Very pale yellow; heavy; curdy
Hg^{2+}	(Mercuric)—bright orange-red; dissolves in excess I^-
Pb^{2+}	Bright yellow
Cu^{2+}	Brown (Cu^{2+} reduced by I^- to Cu^+)

Oxalates; $C_2O_4^{2-}$
Ag^+	White; heavy
Ba^{2+}	White
Ca^{2+}	White
Cu^{2+}	Pale blue
Hg^{2+}	(Mercuric)—white
Ni^{2+}	Very pale green; gelatinous
Pb^{2+}	White

Phosphates; PO_4^{3-}
Ag^+	Light yellow
Ba^{2+}	White; gelatinous
Ca^{2+}	White; gelatinous
Hg^{2+}	White
Pb^{2+}	White
Zn^{2+}	White; gelatinous

Sulfates; SO_4^{2-}
Ba^{2+}	White
Pb^{2+}	White

Sulfides; S^{2-}
Ag^+	Black
Cu^{2+}	Black
Fe^{3+}	(Ferric)—black
Hg^{2+}	Black
Ni^{2+}	Black
Pb^{2+}	Black
Zn^{2+}	White

Sulfites; SO_3^{2-}
Ag^+	White; curdy
Ba^{2+}	White

Note: All Ag^+ precipitates darken eventually on exposure to light.

NAME _____ SECTION _____ DATE _____

(1) Code # of unknowns _____

(2) Description of tests used with each unknown (identify by code #) and observations:

(3) Balanced equations for all reactions used to characterize your unknowns (identify by code #):

(4) Identity of unknowns (identify by code #):

NAME _____ SECTION _____ DATE _____

Answer the following questions before beginning the exercise:

1. The acid species, $H^+(aq)$, may be represented as the *hydronium* ion; that is, the hydrated species. Write an equation showing this association of H^+ with water molecules.

2. In the Discussion, we have this statement: "In the case of salts, the cations are generally simple *aquated* metal ions such as Na^+ . . ." Define the term *aquated* with a diagram, using Na^+ as the cation.

3. Write a *net ionic* equation for the reaction of $AgNO_3(aq) + HCl(aq)$. Show all steps. (Refer to your textbook if necessary.)

24

Acid–Base Titration

OBJECTIVE

To become familiar with a precise analytical technique and to understand the principles of acid–base titration.

DISCUSSION

The concentration of an acid or base is commonly determined by measuring the volume of a base or acid of known concentration required to neutralize it. At the equivalence point, the same number of equivalents of acid and base have been brought together, the number of equivalents having been determined by the volume and concentration of the acid or base. Let us suppose, for example, that 30 mL of 0.05 N (normal) hydrochloric acid are required in the titration of 50 mL of a basic solution of unknown normality. The term *normality* can be defined as the number of milligram equivalents of solute per milliliter (or the number of gram-equivalent weights of solute per liter of solution). The normality of the basic solution can be found as follows: 1 mL of 0.05 N hydrochloric acid contains 0.05 milligram equivalents of HCl. Then 30 mL of the 0.05 N acid contains 30 mL $\times$ 0.05 mg equiv/mL = 1.5 milligram equivalents of acid, and 1.5 milligram equivalents base are required. Because this quantity of base is contained in 50 mL of solution, the normality of the base is

$$\frac{1.5 \text{ mg equiv}}{50 \text{ mL}} = 0.03 \text{ N}$$

In practice, the equivalence point is usually determined by the use of an indicator that changes color at this point. For the titrations in this experiment, phenolphthalein is an excellent indicator, since its color change ("end point") occurs over a pH range compatible with the equivalence point. The reaction being studied is represented by the following generalized equation:

$$HA + BOH \longrightarrow H_2O + BA$$

PROCEDURE

In this exercise you will standardize a solution of a base, NaOH, and use this standarized solution to determine the concentration of acetic acid, a weak acid, in vinegar. Two or more students may cooperate in the exercise, but each individual student should perform an actual titration. Titration is a one-person operation.

1. Setting Up the Titration Assembly

Obtain two clean burets, preferably with Teflon stopcocks. Clamp the burets on a ring stand (Fig. 24.1). If a double buret clamp such as is shown is not available, then use two single buret clamps. Fill both burets with distilled water; then drain each of them through the tip. If droplets of water

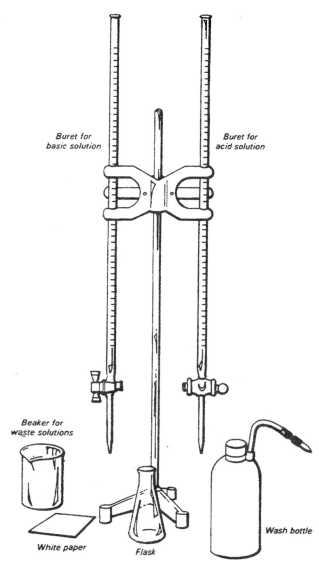

Buret for
basic solution

Buret for
acid solution

Beaker for
waste solutions

Wash bottle

White paper

Flask

Figure 24.1   Titration assembly.

remain on the graduated portion of the buret, it must be cleaned before use (secure directions from the instructor).

2. Preparing a Standard Solution of Sodium Hydroxide

A sodium hydroxide solution cannot be made up directly to an accurately known concentration partly because the solid takes up water during weighing. For these reasons, the solution·is made up approximately to 0.1 N; its exact normality is then determined by a comparison titration with hydrochloric acid of known normality.

Prepare the sodium hydroxide solution by the following method, *if prepared solutions are not available*.

Weigh 4 to 5 g of NaOH pellets (**Protect the eyes.**) into a beaker and dissolve in 100 mL of water. Stir continuously until the solid has dissolved. Transfer the solution to a 1-L beaker (or to a 1-L screw-cap polyethylene bottle if the solution is to be stored), and nearly fill with water. Mix the solution thoroughly and allow to cool. Follow procedure in Section 3.

3. Rinsing and Filling a Buret with NaOH Solution

Remove the buret for base solution from the ring stand and pour in about 5 mL of the NaOH solution. Hold the buret almost horizontal, and rotate in order to rinse the inside surfaces. Drain the solution through the tip into the sink. Rinse again with another 5-mL portion of the base. Finally, fill the buret slightly above the zero mark with the base and clamp it on the ring stand.

4. Standardizing the NaOH Solution

Standardize the NaOH solution according to the following method. Rinse the buret for acid solution twice with 5-mL portions of the standard HCl provided for this experiment and fill to slightly above the zero mark. Drain both burets to slightly below the zero mark and read the level of the bottom of the meniscus (see Part 3G, Fig. G.3) in each buret to two decimal places. Record this and all subsequent data immediately on the Report form (1), (2). It is helpful to hold a blackened card in back of the meniscus and just below to outline its lower surface. Make sure that no bubbles of air are trapped in the tip of the buret. Any bubbles must be removed and the level reread following their removal.

Rinse a clean 250-mL Erlenmeyer flask with distilled water. Add 50 mL of distilled water to the flask and run in acid from the acid buret until the meniscus has fallen to about the 35-mL graduation. Be sure to hold the stopcock in! Add two drops of phenolphthalein indicator. A piece of white paper under the flask will help to make the end point more visible during the titration.

Slowly run in the NaOH solution from the base buret. Swirl the contents of the flask constantly with a rotary motion (Fig. 24.2). Avoid splashing. Wash the sides of the flask occasionally with a fine jet of distilled water from the wash bottle. Near the end point, the trail of color from each drop lasts longer. Approach the end point with care; continue adding base until one drop causes the ap-

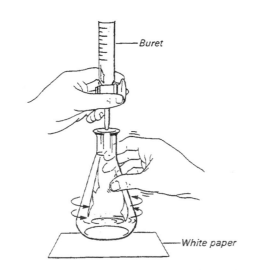

Buret

White paper

Figure 24.2 Titration technique. Swirl the liquid during the titration.

pearance of a pink color that lasts for at least 30 seconds. This is the end point. If a deeper color should result, the end point has been overrun; in this case back-titrate with the acid until one drop causes the color to disappear, and then add $\frac{1}{2}$ drop of the base, or more if necessary. A drop can be split by allowing it to form partially on the buret tip and then washing it down with the wash bottle.

When a satisfactory end point has been attained, read the levels of base (*3*) and acid (*4*). Record the data in the Report form. Record the normality of the standard acid (see label) (*5*). Calculate the normality of the NaOH solution by the procedure given in the Discussion (*6*).

Refill the burets (rinsing is unnecessary) and perform another titration; calculate the normality of the NaOH solution as before (*7*). If the normalities calculated from these two titrations differ by more than 1%, a third titration must be performed. Any error in the standardization of the base will cause an error in the following operation. Use an average of these normalities (*8*) for analyzing the unknown acid in the following section.

5. Analysis of Vinegar

Vinegar is essentially a dilute solution of acetic acid in water; the law requires that vinegar be at least 4% acetic acid by weight. Using a 10-mL pipet, carefully measure 10 mL of vinegar into a 250-mL Erlenmeyer flask; add 100 mL of water and 2 drops of phenolphthalein indicator. Titrate with the standard sodium hydroxide, being careful not to overrun the end point. Titrate two vinegar samples in this way; your results should agree to within 1% in the volume of sodium hydroxide used. Record your data and calculated results (*9*). Include one sample calculation.

The percentage of acetic acid is computed as follows. The volume of sodium hydroxide times its normality equals the number of milligram equivalents of acetic acid neutralized. The equivalent of acetic acid is 60.0 mg, or 0.060 g. The 10-mL sample of vinegar weighs 10.05 g.

Compare the average acetic acid content of your vinegar sample with the legal requirement (*10*).

NAME _____ SECTION _____ DATE _____

4. Standardizing the NaOH Solution

(1)–(8), Data

	Trial 1		Trial 2	
	Base	*Acid*	*Base*	*Acid*
Final buret reading..............	_____ (3)	_____ (4)	_____ (3)	_____ (4)
Initial buret reading	_____ (1)	_____ (2)	_____ (1)	_____ (2)
Volume used.................	_____ mL	_____ mL	_____ mL	_____ mL
Normality of standard acid	_____ (5)		_____ (5)	
Normality of base (calculated)...........	_____ (6)		_____ (7)	
Average normality of base................................. _____ (8)				

5. Analysis of Vinegar (Determining the Concentration of a Weak Acid, Acetic Acid)

(9)

Trial	Volume of standardized base	Percentage of acetic acid*
1		
2		

*Show calculations here:

(10)

213

NAME _____ SECTION _____ DATE _____

Answer the following questions before beginning the exercise:

1. What is meant by a *standard* solution?

2. Define the following terms:
   (a) Molarity

   (b) Normality

3. What is the function of the phenolphthalein used in this acid–base titration?

4. If you overrun the end point in titrating, what can you do to correct this?

5. In titrating, why is it necessary to be sure the stopcock is held in place?

6. As an alternative to using standard HCl to standardize the NaOH solution, a primary standard potassium hydrogen phthalate ($KHC_8H_4O_4$) could be used. If you titrate 1.502 g of this acid with 37.28 mL of NaOH solution, what is the concentration of the NaOH solution? The equation for this standardization is

$$KHC_8H_4O_4 + NaOH \longrightarrow KNaC_8H_4O_4 + H_2O$$

[To determine the concentration of NaOH in the solution, you need to know the number of moles, or equivalents, of NaOH in 0.03728 liter (37.28 mL) of the solution.] Use reverse side to show complete calculation.

25

Electrometric Study of Acid–Base Equilibria

OBJECTIVE

To comprehend acid–base interaction in the context of laboratory operations.

DISCUSSION

Electrometric determination of pH is one of the most widely applied techniques for studying both living and nonliving systems; it far surpasses colorimetric methods in convenience and precision. For electrometric determination, the glass electrode–Ag/AgCl (silver/silver chloride) electrode is most commonly used, although other electrode combinations are possible. The glass–Ag/AgCl combination has the advantages of being usable over a wide pH range, not altering the solution under study, providing direct and precise readings, and being easy to maintain in proper condition.

In aqueous solutions of weak acids the ionization equation is

$$HA \rightleftharpoons H^+ + A^-$$

Where A^- represents the anion of any weak acid—or more accurately,

$$HA + H_2O \rightleftharpoons H_3O^+ + A^-$$

In dilute solutions the law of chemical equilibrium may be applied:

$$\frac{[H_3O^+][A^-]}{[HA][H_2O]} = K_e$$

Because the number of moles of H_2O consumed in the formation of hydronium ions is negligibly small compared to the total number of moles of water in a dilute solution, the above equation may be written

$$\frac{[H_3O^+][A^-]}{[HA]} = K_e[H_2O] = K_i$$

The constant K_i is called an *ionization constant* (see also the Discussion in Exercise 26).

Buffer solutions are mixtures of weak acids and their salts or of weak bases and their salts. They resist a change in hydrogen-ion concentration upon the addition of small amounts of acids or bases.

PROCEDURE

This exercise consists of four sections; your instructor will assign either 1–3, or 4. For Section 4, work in pairs or groups.

1. Ionization Constant of an Acid as Measured in Unbuffered Solutions

Study the instruction manual for the pH meter assigned to you before beginning work with the instrument. Remember that the glass electrode is fragile. Always rinse the electrode with a small portion of the solution next to be investigated before making the determination. Report any difficulties to the instructor.

Standardize the pH meter with the buffer supplied; have the instructor approve any adjustment. Obtain about 50 mL of the unknown acid and record its *exact* concentration (*1*). Using a 10-mL volumetric pipet and a 100-mL volumetric flask, prepare three samples for measurement.

> Sample A: 20 mL of the acid.
> Sample B: 10 mL of the acid + 10 mL of water.
> Sample C: 10 mL of acid diluted to 100 mL.

Record the molarity, measured pH, and calculated K_i for each of these samples in the Report form (*2*). Comment on your results (*3*).

What principle is indicated by your results (*4*)? Suggest some sources of error in this procedure (*5*).

2. Ionization Constant of an Acid as Measured in Buffered Solutions

Obtain about 50 mL of an unknown acid and 50 mL of the sodium salt of the acid. Using 2-mL and 10-mL volumetric pipets and a buret, prepare three sample mixtures as follows (all solutions are 0.1 M unless your instructor specifies otherwise):

> Sample D: 20 mL acid + 4 mL sodium salt of acid + 16 mL NaCl solution.
> Sample E: 20 mL acid + 20 mL sodium salt of acid.
> Sample F: 4 mL acid + 20 mL sodium salt of acid + 16 mL water.

The sodium chloride is added to make the samples equal in ionic strength. Determine the pH of each sample (*6*). Calculate K_i from each reading (*7*), and the average K_i (*8*). Account for any trend in the calculated values of K_i (*9*). Judging from your results, are measurements in buffered and unbuffered solutions equally reliable (*10*)?

3. Buffer Action

Mix 10 mL of 1 M acetic acid, CH_3CO_2H, with 10 mL of 1 M sodium acetate, $Na^+CH_3CO_2^-$. Determine the pH (*11*). Add 5 mL of 0.1 M HCl and note the change in pH (*12*). Compare this pH with the pH of a solution made by adding 5 mL of 0.1 M HCl to 20 mL of water (*13*). What is the mechanism by which buffers stabilize the pH (*14*)?

Determine the effect of adding 5 mL of 0.1 M NaOH to another 20-mL portion of the buffer. Compare the pH change with that obtained by adding 5 mL of 0.1 M NaOH to 20 mL of water (*15*). Account for the difference (*16*).

4. Potentiometric Titration

Obtain access to an apparatus setup as shown in Fig. 25.1. Familiarize yourself with the operation of the components (see instructor) before carrying out the experiment. **Caution: The glass electrode is very fragile; do not bump it against the beaker. Do not use it above pH 12.** See that the stirring bar does not strike the electrodes. Shut off the magnetic stirrer and pH meter before removing the electrode from a solution. If a magnetic stirrer is not available, **the mixture may be very carefully swirled by hand—again taking care not to damage the electrode.**

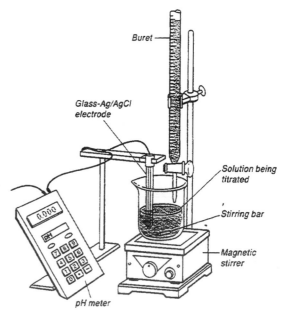

Figure 25.1 Potentiometric determination of pH.

If necessary, standardize the instrument against the pH 7 buffer supplied. The standard NaOH to be used is about 0.2 M; record its exact concentration (*17*). Rinse and fill a clean buret with the base. Record the initial buret reading (*18*). Using a *volumetric* pipet, carefully add 25.00 mL of the unknown acid solution to a beaker, and then add 3 drops of phenolphthalein.

Lower the electrode into the solution and turn on the stirrer (or gently swirl, if stirrer is not available), being careful not to splash. *Stirrer must be off during pH readings*. At first, make pH readings after each addition of 4 mL of base; record the buret reading after each addition of base, and the measured pH (*19*). As you approach the equivalence point, read the pH after increments of 0.2 mL or less. Take two readings after the equivalence point is reached.

Plot pH (vertical axis) against volume of NaOH (horizontal axis) calculated from the recorded data, on graph paper, using the maximum space available (*20*). From the graph, find the pH at the equivalence point (*21*). Also find the pH at half-neutralization (*22*). Calculate concentration of the acid (*23*).

Calculate the ionization constant of the acid from your data (*24*). Report this result to the instructor and compute the percentage of error (*25*). List some of the important sources of error in the determination (*26*).

NAME _____ SECTION _____ DATE _____

Complete the sections assigned by your instructor.

1. Ionization Constant of an Acid as Measured in Unbuffered Solutions

(1) Concentration of unknown acid .... _____ M

(2) Data and Results:

Sample	$[HA]$	pH	Ionization constant, K_i
A			
B			
C			

(3)

(4)

(5) Sources of error:

2. Ionization Constant of an Acid as Measured in Buffered Solutions

(6), (7) Data and Results:

Sample	$[HA]$	$[A^-]$	pH	Ionization constant, K_i
D				
E				
F				

(8) K_i, average .... _____

(9)

(10)

3. Buffer Action

(11)–(13), (15)

Solution	pH
1 M CH_3CO_2H + $NaCH_3CO_2$	
1 M CH_3CO_2H + $NaCH_3CO_2$ + HCl	
0.1 M HCl	
0.1 M CH_3CO_2H + $NaCH_3CO_2$ + NaOH	
0.1 M NaOH	

(14)

(16)

NAME _____ SECTION _____ DATE _____

4. Potentiometric Titration

(17) Concentration of standard NaOH .... _____ M

(18) Initial buret reading .... _____ mL

(19) Titration Data:

Buret reading	Volume of NaOH	pH

(20) Graph of pH (vertical axis) against volume of NaOH (mL) (horizontal axis) (attach to Report).

(21) Equivalence point pH .... _____

(22) Half-neutralization pH .... _____

(23) Calculated concentration of unknown acid .... _____ M

(24) Calculated K_i (show calculations):

(25) Percentage of error (show calculation):

(26) Sources of error:

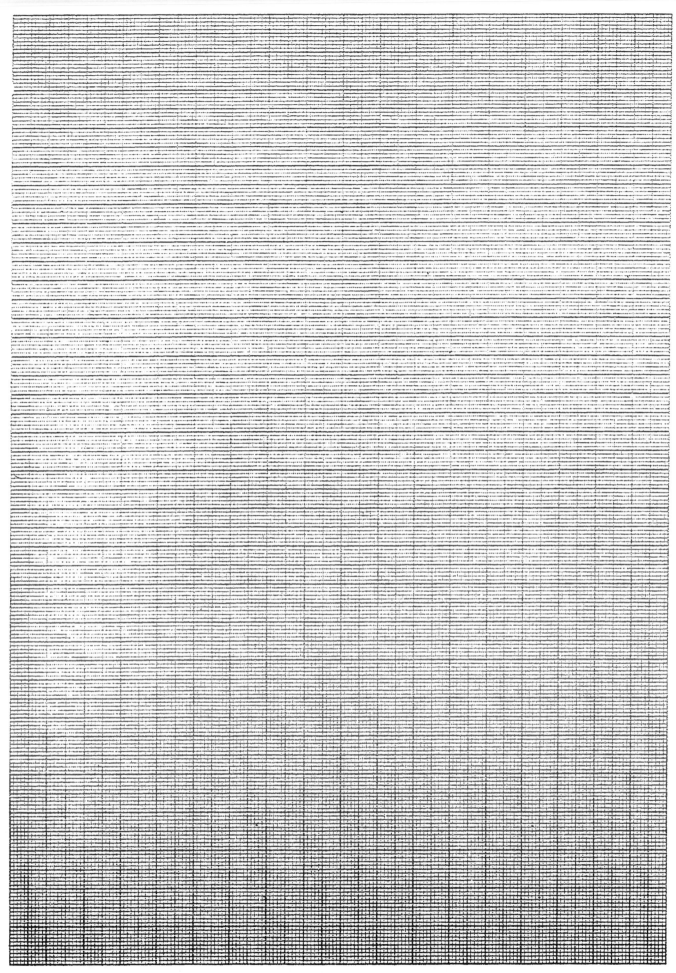

NAME _____ SECTION _____ DATE _____

Answer the following questions before beginning the exercise:

Mathematically, pH $= -\log [H^+]$. What is the pH of 10^{-3} M HCl?

2. What is the $[H^+]$ of a solution that has a pH of 5?

3. Define the term "buffer solution," and give an example of one.

4. Calculate the $[H^+]$ in a 0.10 M acetic acid solution that is 0.10 M with respect to sodium acetate. The K_i constant of acetic acid is 1.8×10^{-5}. What is the corresponding pH?

5. In a potentiometric titration, what is meant by the *equivalence point?*

26

Ionic Equilibria of Weak Electrolytes

OBJECTIVE

To understand the principles governing chemical equilibria involving ions.

DISCUSSION

In solutions of weak electrolytes, ions and un-ionized molecules of the electrolyte are both present; the dissociation of molecules into ions is small. The weak electrolytes include some inorganic acids such as phosphoric acid (H_3PO_4) and carbonic acid (H_2CO_3), most organic acids and bases, and some inorganic hydroxides such as ammonium hydroxide (aqueous ammonia, NH_4OH).

Consider the case of acetic acid (CH_3COOH) equilibrium, which may be shown in the following manner:

$$CH_3CO_2H \rightleftharpoons H^+ + CH_3CO_2^-$$

where $CH_3CO_2^-$ represents the acetate ion. (The participation of water is ignored in this equation since $[H_2O]$ is 55.5 M, and does not change appreciably for dilute solutions.) The equilibrium constant expression is

$$\frac{[H^+][CH_3CO_2^-]}{[CH_3CO_2H]} = K_i = 1.8 \times 10^{-5}$$

For this particular kind of equilibrium, the constant, designated K_i, is called the *ionization constant*. Concentrations of ions and molecules in the equilibrium expression are given in moles per liter. The K_i remains practically the same over a considerable range of concentration, and can be used to determine the ionization of pure acetic acid at all concentrations below one molar; that is, activity corrections are large above this value. For instance, in 0.10 M acetic acid, let C = concentration of H^+ and of $CH_3CO_2^-$. The equilibrium concentration of undissociated acetic acid is $0.10 - C$. Putting these values into the ionization constant expression, we obtain

$$\frac{C \times C}{0.10 - C} = 1.8 \times 10^{-5}$$

Since, as will be seen, $0.10 - C$ is practically equal to 0.10, $C^2 = 0.10 \times 1.8 \times 10^{-5} = 1.8 \times 10^{-6}$; and $C = 1.34 \times 10^{-3}$ mole per liter. Thus, knowing the concentration of H^+ (and of $CH_3CO_2^-$), one can readily calculate the percentage of ionization of acetic acid.

Adding acetate ion to an aqueous solution of acetic acid causes the acetic acid equilibrium, as designated above, to shift to the left as written (refer to Exercise 20).

The resulting decrease in the concentration of H^+ is known as the *common ion effect*. This effect can be observed qualitatively by adding an appropriate indicator to the aqueous solution, and can also be calculated from the ionization constant expression.

When a salt is dissolved in water, the resulting solution may be neutral, basic, or acidic, depending on the nature of the salt. For example, a salt of a strong base and a weak acid, such as sodium acetate, forms aqueous solutions that are basic. The net reaction may be expressed as follows:

$$CH_3CO_2^- + H_2O \rightleftharpoons CH_3CO_2H + OH^-$$

Most chemical reactions possess the quality of reversibility, indicated by equilibrium expressions, and do not "go to completion." Under certain conditions, however, ionic reactions will approach completion.

PROCEDURE

In this exercise you will study qualitatively the equilibria in various solutions of weak electrolytes by means of indicators; you will observe three examples of how ionic reactions approach completion.

1. Equilibria in Solutions of Weak Electrolytes

A. Weak Acids and Weak Bases. To 10 mL of water in a test tube, add 1 or 2 drops of methyl orange solution. (Methyl orange is an acid-base indicator, itself a weak acid, whose equilibrium can be represented by the equation

$$\underset{\text{Red}}{HIn} \rightleftharpoons \underset{\text{Yellow}}{H^+ + In^-}$$

The pH range of this indicator is 3.1–4.4; when the ratio of the In^- to HIn is equal to one, the solution appears orange. The term pH expresses the concentration of hydrogen ion in a solution, which is a measure of the acidity of the solution. This relation expressed mathematically is pH = $-\log [H^+]$.)

Next add a few drops of 0.1 M acetic acid. To what ion is the color change due (*1*)? Dissolve a small amount of solid sodium acetate in the mixture. Explain the resulting color change in terms of the equilibrium between acetic acid and its ions (*2*).

To 10 mL of water in a test tube, add a drop or two of phenolphthalein solution. (The pH range of this indicator is 8.3–10.0, tending from colorless in the acid range to pink in the basic range.)

Next add about 1 mL of 0.1 M aqueous ammonia.

To what is the color change due (*3*)? Now dissolve a small amount of solid ammonium chloride in the mixture. Explain the resulting change in terms of the equilibrium between aqueous ammonia and its ions (*4*).

Repeat the experiment using 0.1 M hydrochloric acid, methyl orange solution, and solid sodium chloride. Account for the result (*5*).

B. Salts. Obtain 5 mL of 1 M sodium chloride, sodium acetate, ammonium chloride, and ammonium acetate. Boil each for 15 seconds to remove dissolved gases. Cool the solutions and test with bromthymol blue indicator by adding 1 drop of the indicator to the 5-mL portions of the solution (*6*). (Bromthymol blue is yellow in acidic solutions and blue in basic solutions; the pH range is 6.0–7.6.) Account for the color change, if any, in each case by means of ionic equations (*7*).

Test a 0.1 M solution of sodium hydrogen sulfate in the same manner as the salts above (*8*). Is the lack of neutrality due, in this instance, to hydrolysis (*9*)? Explain (*10*).

2. Conditions Under Which Ionic Reactions Approach Completion

A. To 5 mL of 0.1 M lead nitrate solution in a test tube add 0.1 M sodium chloride solution dropwise until a precipitate appears. Heat the mixture. Describe and account for the result (*11*). Write the ionic equation representing this equilibrium; include the enthalpy (ΔH) term (*12*).

B. Gradually add 5 mL of 0.1 M hydrochloric acid to 5 mL of 0.1 M sodium carbonate solution in a test tube; describe the result (*13*). Write the ionic equation for the reaction that occurs (*14*).

C. To 10 mL of 0.1 M calcium nitrate solution in a test tube add 10 mL of 0.1 M sodium oxalate solution. The precipitate is calcium oxalate, CaC_2O_4. Decant and discard the liquid, and divide the precipitate into two portions, using two test tubes. To one portion add 5 mL of 0.1 M hydrochloric acid. To the other add 5 mL of 0.1 M acetic acid. Does the precipitate dissolve in both acids (*15*)? Explain the results, using ionic equations; remember that oxalic acid is weaker than hydrochloric acid but stronger than acetic acid (*16*).

Summarize the conditions under which ionic reactions go to completion as observed in **A**, **B**, and **C** (*17*).

1. Equilibria in Solutions of Weak Electrolytes

    A.

(1)

(2)

(3)

(4)

(5)

    B.

(6)

(7) Ionic equations:

(8)

(9)

(10)

2. Conditions Under Which Ionic Reactions Approach Completion

A.

(11)

(12) Ionic equation:

B.

(13)

(14) Ionic equation:

C.

(15)

(16) Ionic equations:

(17)

NAME _____ SECTION _____ DATE _____

Answer the following questions before beginning the exercise:

1. In the Discussion it is stated that the participation of water is ignored in the acetic acid equilibrium since $[H_2O]$ is 55.5 M. Show how this value, 55.5 M, is determined.

2. Rewrite the equation for the acetic acid equilibrium *including* the participation of water. Name the ions produced.

3. Define the term "common ion effect," and give an example of it.

4. What is meant by *hydrolysis?*

5. Give an example of
   (a) a neutral salt:

   (b) a basic salt:

   (c) an acidic salt:

6. A net chemical change is observed when ionic reactions approach completion. Refer to your textbook, and list three conditions under which ionic reactions approach completion.

Oxidation and Reduction

OBJECTIVE

To apply the principle of oxidation–reduction to constructing a voltaic cell.

DISCUSSION

An element is said to have undergone oxidation in a reaction if its oxidation state has increased. In the reaction $2Mg + O_2 \rightarrow 2MgO$, the oxidation state of magnesium has increased from zero to $+2$, a process of *oxidation*. The element that causes oxidation must itself decrease in oxidation state; in the example given, the oxidation state of oxygen has changed from zero to -2. Such a process is termed *reduction*. Invariably, oxidation and reduction occur simultaneously.

Theoretically, oxidation–reduction reactions involve transfer of electrons from reducing agent to oxidizing agent; in the example cited, electrons pass from magnesium to oxygen. Electron transfer can be demonstrated by placing suitable reactants in separate vessels connected by a "salt bridge," with each vessel containing an immersed electrode connected by means of a metallic wire; the spontaneous flow of electricity can be observed when a suitable voltmeter or light bulb is included in the circuit.

The electromotive force (emf) value for a half-reaction changes when the concentration changes. The Nernst equation makes it possible to calculate emf values at other concentrations. The Nernst equation for 25°C is

$$E = E° - \frac{0.059}{n} \log Q$$

where E = emf for the reaction
$E°$ = standard electrode potential*
n = number of electrons stated in the half-reaction
Q = an expression that takes the same form as the equilibrium constant for the half-reaction without indicating the electrons

For example, if we use the Nernst equation with the half-reaction

$$Fe^{3+} + e \longrightarrow Fe^{2+}$$

then

$$E = 0.77 - \frac{0.059}{1} \log \frac{[Fe^{2+}]}{[Fe^{3+}]}$$

---

*Standard electrode (reduction) potentials are obtained from tables in a handbook. They refer to 1 M solutions of ions, a pressure of 1 atm for gases, and a temperature of 25°C.

To obtain the net cell reaction (net emf of the cell), one adds the half-reactions. When the resultant emf of the cell is positive, the reaction proceeds spontaneously to the right, as written.

PROCEDURE

In this exercise you will construct a voltaic cell based on an oxidation–reduction system.

Clamp two 200 × 25-mm test tubes side by side. Attach two electrodes (stainless steel, platinum, or nichrome) to a high-resistance voltmeter (V) with copper wire and suspend them in the test tubes, as shown in Fig. 27.1.

Nearly fill one test tube with 0.1 M iron(II) ammonium sulfate solution acidified with sulfuric acid and the other test tube with 0.02 M potassium permanganate solution (prepare from 0.1 M) containing 1 mL of 3 M sulfuric acid.

Prepare a salt bridge by bending a 12-cm length of 8-mm glass tubing in the form of a U (see Part 3C, page 388); fill it completely with a saturated solution of sodium sulfate. Plug the ends with firm masses of cotton that have been soaked in the sodium sulfate solution; eliminate all air bubbles. Invert the bridge in the test tubes, being careful not to trap any air.

Read the voltage on the voltmeter set on the 0–10 volt range (preferably a vacuum tube voltmeter) and record the reading; reverse connections if necessary to get a reading (1). Which electrode is positive (2)? Account for this in terms of tendency for electron flow (3).

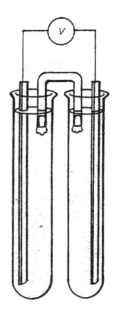

Figure 27.1   A simple voltaic cell.

Allow the cell to operate by removing the voltmeter and connecting the electrodes with a wire. Note any changes that appear in the first few minutes (4). Observe again at the end of the laboratory period and, if possible, also at the beginning of the next period (5). Write the equations for the reaction occurring at each electrode and their summation (6). Diagram the flow of electrons and ions in the system (7).

NAME _____ SECTION _____ DATE _____

(1) Volts:

(2) Positive electrode:

(3)

(4)

(5)

(6) Reaction at anode:

   Reaction at cathode:

   Summation:

(7) Flow diagram of electrons and ions:

NAME _____ SECTION _____ DATE _____

Answer the following questions before beginning the exercise:

1. List the factors that determine the electromotive force (emf) of a voltaic cell.

2. Referring to tables of standard electrode reduction potentials, calculate the standard emf of an iron(II) versus MnO_4^- cell.

3. Apply the Nernst equation to correct the emf found in Question 2 for the concentrations used.

241

Determination of Iron by Permanganate Titration

OBJECTIVE

To become familiar with a precise analytical technique involving oxidation–reduction.

DISCUSSION

The gram equivalent weight of a substance involved in a given reaction can be defined as the number of grams of the substance associated with the transfer or neutralization of N electrons or protons or N negative or positive charges, where N is Avogadro's number, 6.022×10^{23}. An oxidation–reduction reaction between inorganic ions involves a transfer of electrons from the substance oxidized to the substance reduced. In a given reaction, the weight of the substance yielding that quantity of the ion that gains or loses N electrons is the gram equivalent. For example, in acid solution, potassium permanganate, $KMnO_4$, is reduced to manganese(II) ion, Mn^{2+}; this amounts to a gain of five electrons for the manganese atom. Consequently, the equivalent weight of $KMnO_4$ is one-fifth its formula weight, or 31.6 grams. Potassium permanganate cannot be obtained in a state of high purity; therefore, its standard solution is made up approximately by weight and then standardized against some reducing agent of known purity, such as pure sodium oxalate or specially prepared iron wire. Because one gram equivalent of an oxidizing agent reacts with exactly one gram equivalent of a reducing agent, the results of oxidation–reduction titrations can be calculated in the same way that is described for acids and bases in Exercise 24.

PROCEDURE

In this exercise you will prepare or be provided with a potassium permanganate solution. (*Note:* If you are to prepare your own solution, it must stand overnight.) This solution you will standardize against a standard sodium oxalate solution by titration; you will then use the standardized potassium permanganate solution to determine the amount of iron in an unknown sample.

1. Preparation and Standardization of the Potassium Permanganate Solution

Use the general technique described in Exercise 24 for the preliminary operations. Weigh about 3.2 g of potassium permanganate crystals, and dissolve them in 1 L of water. Allow the solution to stand in a dark bottle overnight. Filter off any solid that accumulates.

NAME _____ SECTION _____ DATE _____

1. Preparation and Standardization of the Potassium Permanganate Solution

	Trial 1 (2)–(5)	*Trial 2* (6), (7)
(1) Normality of oxalate	_____	_____

	$KMnO_4$	$Na_2C_2O_4$	$KMnO_4$	$Na_2C_2O_4$
Final buret readings	_____ mL	_____ mL	_____ mL	_____ mL
Initial buret readings	_____ mL	_____ mL	_____ mL	_____ mL
Volume	_____ mL	_____ mL	_____ mL	_____ mL

Normality of permanganate (5) _____   (7) _____

(8) Average of (5) and (7) .... _____

2. Determination of Iron in an Unknown Sample

	Trial 1 (9)–(11)	*Trial 2* (12)
Weight of unknown	_____ g	_____ g
Final buret reading	_____ mL	_____ mL
Initial buret reading	_____ mL	_____ mL
Volume of standardized permanganate......................	_____ mL	_____ mL
Fe in unknown (show calculation)	_____ mg	_____ mg

(13) Average.................................... _____

NAME _____ SECTION _____ DATE _____

Answer the following questions before beginning the exercise:

1. The overall reaction of permanganate with sodium oxalate in acid solution is given by the following equations:

$$2Na^+ + C_2O_4^{2-} + 2H^+ \longrightarrow H_2C_2O_4 + 2Na^+$$

$$2MnO_4^- + 5H_2C_2O_4 + 6H^+ \longrightarrow 2Mn^{2+} + 10CO_2 + 8H_2O$$

   (a) What substance is oxidized?

   (b) What substance is reduced?

   (c) What is the oxidation number of Mn in MnO_4^-?

   (d) What is the oxidation number of C in $H_2C_2O_4$?

2. Write the equation for the reaction of MnO_4^- with Fe^{2+} ion. (The reaction is carried out in acid solution.)

3. Given the following data, calculate the percentage of iron in 2.00 g of an unknown sample: 32.0 mL of 0.100 N permanganate were used to titrate the iron sample. Each milligram equivalent of iron weighs 558.5 mg.

29

Electrochemical Cells

OBJECTIVE

To comprehend and apply principles governing the interaction of free electrons with molecules and ions in solution.

DISCUSSION

If two substances capable of reaction by electron transfer are placed in separate vessels, reaction may still occur under certain conditions. First, a pathway between the two solutions must be provided for the electrons; a metal wire forms such a pathway. Second, provision must be made for avoiding local accumulation of electric charge resulting from the transfer of electrons; a liquid path performs this function by permitting a compensatory migration of ions. The complete device is termed a cell. Two types of cells can be devised—*voltaic* and *electrolytic*. The former operates spontaneously, producing electrical energy and generating a voltage that depends partly on the particular reaction involved (see Exercise 27). Electrolytic cells, on the other hand, must be operated by electrical energy supplied from an outside source. Many important commercial processes depend on electrolytic cells; among them are aluminum, magnesium, and chlorine production.

One diagram of a simple voltaic cell is shown in Fig. 27.1 (Exercise 27). In this case the liquid path permitting migration of *ions* between the two solutions is a salt bridge. Another design of a voltaic cell is shown in Fig. 29.1. In this case the liquid path permitting the migration of *ions* between the two solutions is a porous barrier.

The salt bridge or porous barrier electrically connects the two half-cells without allowing the two solutions to mix appreciably. The *flow of electrons*, in either case, proceeds by way of the electrodes and metal wire.

An electrolytic cell is shown in Fig. 29.2.

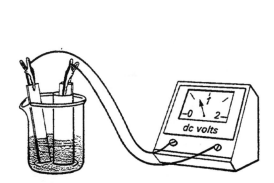

Figure 29.1 A voltaic cell.

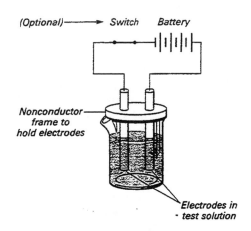

(Optional) ⟶ Switch Battery

Nonconductor frame to hold electrodes

Electrodes in test solution

Figure 29.2 An electrolytic cell.

In electrochemical cells, distinction between anode and cathode is made by the chemical reaction taking place there:

Anode: oxidation takes place (electrons given up to the external circuit)
Cathode: reduction takes place (ions reacting with electrons from the external circuit)

The anode is to be attached to the "minus" or "reference" lead from the voltmeter. The cathode is to be attached to the "plus" or "input" lead from the voltmeter. With this convention, the measured voltage (emf), including sign, is the value of the emf of the complete cell. If the voltage is plus, the reaction is spontaneous in the forward direction; if negative, the reaction is spontaneous in the reverse direction.

Michael Faraday found that the quantity of substances that are undergoing change at each electrode during electrolysis is directly proportional to the quantity of electricity that passes through the electrolytic cell. Faraday's law can be stated in quantitative terms: During electrolysis, 96,487 coulombs, or 1 faraday, of electricity (6.022×10^{23} electrons) reduce one gram-equivalent of the oxidizing agent and oxidize one gram-equivalent of the reducing agent. (A coulomb is that quantity of electricity involved when a current of one ampere flows for one second.)

PROCEDURE

You will perform Sections 1 and 2 of this exercise in groups, as assigned by your instructor.

1. A Voltaic Cell

Clean 1 × 3-cm strips of copper and zinc and attach a 30-cm copper wire to each through a small hole in the strip, or by means of alligator clips (see Fig. 29.1). Place 40 mL of 1 M copper(II) sulfate in a 150-mL beaker and support in this solution a small cup of porous porcelain (or an F or VF sintered-glass filtering crucible) nearly filled with 1 M zinc sulfate; do not allow the two solutions to mix.

Place the zinc strip in the solution of zinc sulfate and the copper strip in the solution of copper(II) sulfate. Connect the cell to a voltmeter (see Discussion) and measure the voltage (emf) of the cell (1).

Compare this reading with the voltage (emf) calculated from the half-reaction potentials of copper and zinc, obtained from tables in the text or a handbook (2). Account for any difference between calculated and measured values (3).

Repeat the experiment, using 40 mL of 2 M copper(II) sulfate in a clean 150-mL beaker, and 2 M zinc sulfate in a rinsed porous cup. Measure the voltage (4). What effect does concentration have on voltage delivered (5)?

Repeat the experiment above, but this time cut the electrodes in half; that is, use 0.5 × 3-cm strips of copper and zinc. Measure the voltage (6). What effect does the size of the electrode have on voltage delivered (7)?

Draw a diagram of this cell showing both the electron flow and the ion flow (8).

2. Electrolytic Cells

A. In a small beaker, mix 1 mL of 0.1 M sodium iodide solution, 9 mL of water, 1 mL of fresh starch solution, and 2 drops of phenolphthalein indicator. Lead wires from a dry cell or storage battery into this solution and observe the region around each wire (9). Write half-reaction equations for the reactions (10). What would happen if the reaction products were to come into contact (11)? Why (12)?

B. Polish a 1 × 3-cm copper strip until it is very bright and attach to it a wire leading to the negative pole of a 6-V dc source. Fasten another wire to a small carbon electrode and lead this wire to the positive pole of the dc source (see Fig. 29.2). Place the electrodes in a 50-mL beaker and add 35 mL of the "nickel solution for electroplating."* Allow the action to proceed until a good coating is

---

*To prepare, use 90 g of $NiSO_4 \cdot 6H_2O$, 30 g of $NiCl_2 \cdot 6H_2O$, 15 g of boric acid, and 1 L of water.

obtained on the copper strip. Record your observations (*13*).

Disconnect the wires and wash the plated electrode thoroughly; finally, polish it with a clean cloth. Describe the deposit (*14*).

Write equations for the half-reactions, assuming that hydrogen is evolved at the positive electrode (*15*). Suggest some conditions to be maintained during electroplating to avoid a coating that is porous and nonadherent (*16*).

NAME _____ SECTION _____ DATE _____

1. A Voltaic Cell

(1) Cell voltage (emf), measured .... _____ V

(2) Cell voltage (emf), calculated:

$$Zn \; \rightleftharpoons \; Zn^{2+} + 2e \; \underline{\hspace{1cm}} \; V$$

$$Cu^{2+} + 2e \; \rightleftharpoons \; Cu \; \underline{\hspace{1cm}} \; V$$

_____

$$Zn + Cu^{2+} \; \rightleftharpoons \; Zn^{2+} + Cu \; \underline{\hspace{1cm}} \; V \; (emf \; of \; cell)$$

(3) Difference between (1) and (2): _____ V
Reason for difference:

(4) Cell voltage (emf), measured .... _____ V

(5) Effect of concentration on voltage:

(6) Cell voltage (emf), measured .... _____ V

(7) Effect of size of electrode on voltage:

(8) Diagram of cell, showing both electron flow and ion flow:

2. Electrolytic Cells

A.

(9) Appearance near anode:

Appearance near cathode:

NAME _____ SECTION _____ DATE _____

Answer the following questions before beginning the exercise:

1. Write the half-reactions for the following oxidation–reduction reaction:

$$Cd(s) + 2Ag^+(aq) \longrightarrow 2Ag(s) + Cd^{2+}(aq)$$

2. Describe a voltaic cell made to study the reaction shown in Question 1.

3. Calculate the emf of this cell from the half-reactions of Cd and Ag (see text or handbook).

4. What type of cell is an automobile battery? Explain.

5. Using reference materials, explain how a flashlight battery operates.

6. Using Faraday's law (see the Discussion), calculate the mass of nickel that would be deposited by a current of 0.1 ampere flowing for 10 hours, and by a current of 10 amperes flowing for 1 hour.

7. Would equally good coatings be obtained in both of the cases referred to in Question 6? Explain.

Determination of the Acidity of Natural Waters

OBJECTIVE

To become familiar with titration using indicators and the pH meter as methods of analyzing water for acidity; to understand that the acidity of water may be related to pollution.

DISCUSSION

All natural waters are impure since they contain dissolved substances whose nature depends on the kind of rocks and soil in contact with the water. In addition to these natural contaminants in water, many pollutants are added by human beings. *Pollutants* are residues of things we make, use, and discard. The effect of these pollutants on the environment is an area of concern.

There are two senses, actually parts of the same system, in which acidity enters into water quality: (1) total "reserve" acidity, and (2) instantaneous acidity. Total "reserve" acidity is determined by titrating the H_2CO_3, HCO_3^-, H_3PO_4, and organic acids present in water. Instantaneous acidity is the hydrogen ion concentration of the water sample measured by a pH meter. The equilibrium existing between H_2CO_3, HCO_3^-, and CO_3^{2-} in natural waters is disturbed if acid pollutants enter; the extent of the disturbance is estimated from the difference in NaOH required to reach pH 4.5 (the methyl orange end point) and to reach pH 8.3 (the phenolphthalein end point). At pH 4.5 only strong acids have been neutralized; at pH 8.3 these and weak acids and acid salts have been neutralized.

Acid–base indicators (such as methyl orange and phenolphthalein) used in titration are sensitive to changes in the hydrogen ion concentration of a solution. The concentration of the hydrogen ion is a measure of the acidity or basicity of a solution. It has been found convenient to express the concentration of the hydrogen ion in terms of the negative logarithm of the hydrogen ion concentration. This is the pH of the solution. Thus

$$pH = -\log [H^+] = \log \frac{1}{[H^+]}$$

The pH at which an indicator changes color depends on the particular indicator; methyl orange indicator changes color (red to orange) at pH 4.5 and phenolphthalein indicator changes color (colorless to pink) at pH 8.3. Thus a water sample in which methyl orange is yellow and phenolphthalein is colorless has a pH range of 4–9, the range of pH for natural waters.

The pH meter is a direct readout instrument, employing a glass electrode. The meter is standardized before use with a buffer solution of a known pH.

The range of pH for natural waters is 4–9. Usually the waters are slightly basic because of CO_3^{2-} and HCO_3^- present, but pollutants, such as industrial wastes, may alter the pH considerably. Indicators cannot be used to determine pH of natural waters, since they are affected by many constituents of water and also may themselves induce a change in the water. The pH meter employing a glass electrode is preferable and can be used to monitor a flowing stream.

PROCEDURE

In this exercise you will determine the acidity of a sample of natural water, and assess its quality on the basis of your results.

1. Total "Reserve" Acidity

Obtain a sample of the natural water. Remove any chlorine by adding a drop of 0.1 N sodium thiosulfate. Add the volume the instructor specifies (*1*) to a 250-mL Erlenmeyer flask; 50–100 mL of water are usually satisfactory.

Obtain a buret, rinse with standard 0.02 N NaOH, and fill. Place a test tube over the top of the buret to prevent excessive CO_2 absorption. Record the concentration of the base (*2*).

Place the flask of water being tested on a sheet of white paper and add 2 drops of methyl orange indicator. Using the technique of Exercise 24, titrate to a faint orange end point. Record the volume of NaOH used (*3*).

Obtain another sample of the same natural water. Following the same procedure, titrate to the phenolphthalein end point. Record the volume of NaOH used (*4*).

Express the results as parts per million of $CaCO_3$ (ppm = mg/liter). Example:

10 mL of 0.02 N NaOH = 0.2 meq/mL
$$2NaOH + 2H^+ \rightarrow 2Na^+ + H_2O$$
1 meq/mL Na^+ = 0.5 meq/mL $CaCO_3$
0.2 meq/mL Na^+ = 0.1 meq/mL $CaCO_3$
0.1 meq/mL $CaCO_3$ = 10 mg/mL $CaCO_3$
  or
10,000 mg/L = 10^4 ppm

Report ppm $CaCO_3$ at pH 4.5 (*5*) and at pH 8.3 (*6*). Calcium carbonate usually ranges from 10 to 500 ppm.

Summarize your conclusions about the water tested (*7*).

2. Instantaneous Acidity

Caution: The glass electrode is very fragile, and must not be allowed to "dry" out. See your instructor for details of operation for the pH meter available to you.

Standardize a pH meter against the buffer supplied.

Determine the pH of several samples of natural waters from different sources. Record the type of sample (*8*), its source (*9*), and the pH (*10*).

Try to account for the differences observed (*11*).

NAME _____ SECTION _____ DATE _____

1. Total "Reserve" Acidity

(1) Volume of H_2O .... _____ mL

(2) Concentration of base, NaOH .... _____ N

(3) Volume of NaOH (methyl orange) .... _____ mL

(4) Volume of NaOH (phenolphthalein) .... _____ mL

(5) ppm $CaCO_3$ at pH 4.5 .... _____

(6) ppm $CaCO_3$ at pH 8.3 .... _____

(7) Summary:

2. Instantaneous Acidity

(8) Sample:

(9) Source of sample:

(10) pH:

(11)

NAME _____ SECTION _____ DATE _____

Answer the following questions before beginning the exercise:

1. What precautions must be taken when using a pH meter with a glass electrode?

2. How do you standardize a pH meter?

3. The $[H^+]$ of a solution is 10^{-8} mol/l. What is its pH?

4. The pH range is from 1–14. What is neutral? What is acidic? What is basic (alkaline)?

5. In general, what is the effect of pollution on the pH of natural waters?

6. Show the equilibrium of the dissociation of H_2CO_3:

31

Determination of the Hardness of Water

OBJECTIVE

To employ a titration method using an indicator and a chelating agent for the determination of "hardness" of water.

DISCUSSION

"Hardness" of water is due principally to the presence of the ions Ca^{2+} and Mg^{2+}. These ions form precipitates with the anions of soaps, wasting soap and contaminating surfaces. At high temperatures, some hard waters also deposit scale in boilers and piping systems. Scale consists of insoluble substances such as calcium carbonate, magnesium carbonate, iron(II) carbonate, and calcium sulfate. Much of the scale is calcium sulfate, which is *less* soluble in hot water than in cold and precipitates partly because of that fact. Hardness, usually reported as $CaCO_3$, ranges up to several hundred parts per million.

The standard method for determining water hardness involves titrating the sample with ethylenediaminetetraacetic acid, EDTA. This reagent, a chelating agent, complexes the ions Ca^{2+} and Mg^{2+}. When practically all free Ca^{2+} and Mg^{2+} have been complexed (essentially removed from solution), the indicator, eriochrome black T, presents changes in color from red to blue, signaling the equivalence point in the titration.

PROCEDURE

In this exercise you will determine the hardness of a sample of water supplied, and assess its quality on the basis of your results.

A *buffer*, a mixture that resists a change in hydrogen ion concentration, is added before the titration.

Measure 50.0 mL of hard water, or the volume of water the instructor designates, into an Erlenmeyer flask, add 1 mL of the supplied buffer, and 4 drops of the indicator. Record the volume of water used (*1*) and its source (*2*).

If necessary, review the details of the method of titration (see Exercise 24), before proceeding.

Rinse and fill a buret with the standard EDTA solution supplied and record its normality (*3*). Titrate carefully with EDTA until the initial red color disappears and a pure blue color appears. Record the volume of EDTA used (*4*).

Run two more trials of this titration in the same manner, starting out with 50.0-mL (or other designated volume) samples of the same hard water provided. Record the data as before in the Report form.

Now calculate the mean (average) value of the volume of EDTA used in the three trials (*5*).

Calculate the precision (standard deviation) of your data (see the Appendix) (*6*).

263

Calculate and report the total hardness of the water sample as parts per million (ppm) of $CaCO_3$ (7); this is the same as milligrams (mg) $CaCO_3$ per liter. The volume of EDTA times its normality is taken to be the number of mg equivalents of $CaCO_3$ in a 50.0-mL sample of water.

What is your conclusion about the hardness of the sample of water you analyzed (8)?

NAME _____ SECTION _____ DATE _____

	Trial 1		Trial 2		Trial 3	

(1) Volume of H_2O .... _____ mL   _____ mL   _____ mL

(2) Source of sample:

(3) Normality of EDTA .... _____ $\dfrac{\text{mg equiv}}{\text{mL}}$

(4) Volume of EDTA .... _____ mL   _____ mL   _____ mL

(5) Mean (average) volume of
    EDTA .... _____ mL

(6) Precision of data (standard deviation) (show calculation):

(7) ppm $CaCO_3$ (that is, mg $CaCO_3$ per liter) (show calculation, based on your data):

(8)

1. Explain what is meant by "hard" water.

2. What is a *chelating* agent?

3. Why is "hard" water undesirable for home use?

4. What is unique about the solubility of $CaSO_4$ in *hot* water? How does this affect water usage?

32

Spectrophotometric Determination of Phosphate, a Water Pollutant

OBJECTIVE

To become familiar with a colorimetric technique using a spectrophotometer, for analyzing water.

DISCUSSION

Among the advantages of measuring color intensity by electronic means are relative speed, sensitivity, and accuracy. Visual comparison of color intensity is complicated by human errors such as fatigue and the need for maintaining color standards over long periods.

Typically one uses a *spectrophotometer*, selecting a wavelength of incident radiation corresponding to a region of strong absorption, to measure the degree of *absorption* for a given sample at this wavelength. (In Exercise 10, the wavelengths associated with *emission* of radiation were observed.) If the constituent in the sample under analysis has no color, it may be changed chemically into a colored substance. The phosphate ion is colorless. When it is treated in acid solution with ammonium molybdate, molybdophosphoric acid forms; this is then reduced to molybdenum blue, a coordinated metal complex, with the combination of aminonaptholsulfonic acid and sulfite agents. Molybdenum blue absorbs strongly at the wavelength of 690 nm.

Phosphate occurs in traces in many natural waters. Often it occurs in significant amounts, for instance, during periods of low biologic productivity. The growth rate of algae increases in the presence of phosphate, and the algae tend to deplete the store of oxygen in the water. Raw or treated sewage, agricultural drainage, and industrial effluents often contain large concentrations of phosphate.

Most spectrophotometers have both a linear % transmittance scale and a nonlinear (logarithmic) absorbance scale. *Transmittance* (T) as read on an adjusted meter is that fraction of radiant power passed through the sample:

$$T = \frac{P}{P_0} \left(\% \, T = \frac{P}{P_0} \times 100 \right)$$

where P_0 is the radiant emergent from the reference or blank solution. The use of the blank compensates for all losses of absorption except the sample measured.

Absorbance (*A*), related to transmittance, is represented by the following derived equation:

$$A = \log \frac{P_0}{P} = abc$$

where *a* is a constant, *b* is the cell path of the instrument, and *c* is the concentration of the absorbing sample. This relationship is summarized below:

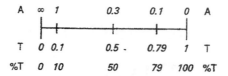

PROCEDURE

In this exercise you will carry out a colorimetric determination of the pollutant phosphate ion in natural water samples.

The use of a Bausch and Lomb Spectronic 20 spectrophotometer (see Fig. 32.1) equipped with an accessory filter and phototube that increases the spectral range of the instrument to 950 nm is outlined. (If another type of spectrophotometer, such as a Spectronic 21, is to be used, see your instructor for details of its operation.) The reading on the sample of water analyzed is evaluated by reference to a previously determined standard calibration curve. To prepare the standard curve, measure the % transmittance of each of the standard phosphate ion solutions after proper treatment of them as described below, and record the data (*1*). Plot % transmittance against phosphate ion concentration on the graph paper provided (*2*). (For an optional graphing method, see suggestion found in Exercise 14, page 126.)

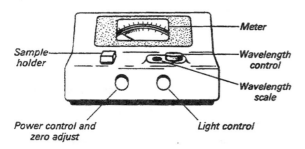

Figure 32.1 Bausch and Lomb Spectronic 20 spectrophotometer.

1. Preparation of Standard Curve

(All solutions used should be at the same temperature, between 20°C and 30°C.)

A. Clean and dry six 125-mL Erlenmeyer flasks. Label them 5, 10, 15, 20, 25, and 30. Make a set of six standard phosphate ion solutions, ranging from 5 mg per liter to 30 mg per liter; that is, 5, 10, 15, 20, 25, and 30 mg per liter, by placing about 75 mL of each in its appropriate labeled flask.

Have six 13 × 100-mm clean, dry test tubes (or Spectronic 20 cuvettes, if available) ready nearby.

Prepare a proper blank; that is, a solution that contains all the components of the mixtures to be analyzed *except for phosphate ion:* 50.0 mL distilled water, 2.0 mL of the molybdate solution, and 2.0 mL of sulfonic acid reagent. Setting the machine with the blank ensures that the % transmittance of the standard sample to be measured is due only to the phosphate complex.

B. Familiarize yourself with the operation of the spectrophotometer to be used. Operating instructions for the Spectronic 20 (Fig. 32.1) follow:

1. Turn on the Spectronic 20 by turning the left knob clockwise until a click is heard. A pilot light indicates when the meter is on. Allow the instrument to warm up for 15 minutes.
2. Adjust the wavelength to 690 nm, using the knob to the right of the sample chamber.
3. Cover the empty sample chamber, adjust the meter to read 0% transmittance, top scale on meter, using the left knob.
4. Fill a clean, dry 13 × 100-mm test tube (or cuvette) with the proper blank you have prepared. Wipe it clean and place it in the sample chamber.
5. Now adjust the meter to 100% transmittance, using the lower right knob.
6. Remove the blank, and leave the settings you have determined.

The instrument is now adjusted for measurement of the phosphate ion samples. Periodically check the 0% and 100% T settings to make sure they have not drifted. The 100% T is checked by reinserting the sample tube containing the blank.

C. Treat *each* standard phosphate ion solution (from Part A) separately as follows:

1. Using a clean, dry 50-mL graduated cylinder, or preferably a 50-mL pipet, measure out 50.0 mL of the first phosphate ion sample into another clean, dry 125-mL flask.
2. Using a 10-mL graduated cylinder, add 2.0 mL of the molybdate solution provided and swirl.
3. BE FOREWARNED that the spectrophotometer reading must be made *exactly 5 minutes* after adding and swirling the next reagent; namely, 2.0 mL of sulfonic acid reagent. If the instrument is available and is adjusted as described in Part B above, then proceed.
4. Pour enough of your treated phosphate ion sample into one of your clean, dry 13 × 100-mm test tubes, place in the sample chamber, and measure its % transmittance. Record the result in the Report form.

Continue in the same manner with the remaining five standard phosphate ion solutions, each time using a clean, dry graduated cylinder and Erlenmeyer flask.

2. Analysis of Water Sample

Obtain a *clear* natural water sample that has a pH between 4 and 10, and contains not over 1.5 mg of phosphate ion (30 mg per liter). [If a dilution of the sample is made, account for this in (3) of the Report.]

Measure out 50.0 mL of the water sample into a clean, dry 125-mL Erlenmeyer flask.

Treat exactly as the standard phosphate ion solutions were treated (see Section 1 above). Be sure the spectrophotometer has remained adjusted.

Locate the % transmittance of the natural water phosphate ion sample on the standard curve and read off the concentration (3). Express as parts per million (ppm) PO_4^{3-} (4).

Repeat the determination and record the result in the Report form. Average the results (5). Comment on their agreement (6).

NAME _____ SECTION _____ DATE _____

1. Preparation of Standard Curve

(1)

Flask	Phosphate ion concentration, mg/L	% Transmittance
5	5	_____
10	10	_____
15	15	_____
20	20	_____
25	25	_____
30	30	_____

(2) Standard curve: Plot of % transmittance against phosphate ion concentration
   on graph paper provided; attach to Report.

2. Analysis of Water Sample

	Trial 1	Trial 2
(3) Percent transmittance of natural water phosphate sample	_____	_____
Concentration of phosphate ion, mg/L	_____	_____
(4) Expression of (3) as ppm PO_4^{3-}:	_____	_____

(5) Average of (4):

(6)

NAME _____ SECTION _____ DATE _____

Answer the following questions before beginning the exercise:

1. Why must molybdate and sulfonic acid reagents be added to the standard
    phosphate ion solutions before measuring them spectrophotometrically?

2. What is the name of the complex formed when the samples are treated as in
    Question 1?

3. If one did not know at what wavelength the complex absorbs, how could this
    be determined experimentally?

4. How would a plot of absorbance against concentration differ from a plot of
    % transmittance against concentration?

277

5. Why is phosphate ion considered to be a *pollutant* in natural waters?

6. Why do you think the color must be measured exactly 5 minutes after developing, in each case?

7. Suppose one were measuring the % transmittance of solutions of $Co(NO_3)_2$. What would be a suitable blank to use in adjusting the spectrophotometer? Give reasons for your answer.

8. List important sources of possible error in this method.

33

Properties of Oxygen and Hydrogen

OBJECTIVE

To become familiar with some of the properties of oxygen (O_2); to become familiar with some of the properties of hydrogen (H_2).

DISCUSSION

A source of supply of oxygen and hydrogen is assumed for purposes of this exercise. Common laboratory methods for the preparation of these gases are given in the Appendix.

Oxygen is the most abundant of the earth's elements; it is combined with various elements in the earth's crust and in the bodies of plants and animals, with hydrogen in water, and exists in the molecular form (primarily O_2 and some O_3, ozone) in the atmosphere. Oxygen (O_2) is a colorless, odorless, and tasteless gas at ordinary temperatures. It is slightly more dense than air. Oxygen is only slightly soluble in water, but this solubility is vital to aquatic life. Oxygen is noted for its chemical activity; it forms compounds either directly or indirectly with nearly all of the other elements. The chemical activity of oxygen is greatest at higher temperatures, when pure, and in the presence of a catalyst. The oxides (compounds containing oxygen) of fairly active metals react with water to form basic hydroxides; for this reason such oxides are called *basic oxides* or *basic anhydrides*. These solutions turn red litmus blue, a property of aqueous solutions of bases. On the other hand, the oxides of many nonmetallic or electronegative elements are termed *acidic oxides* or *acidic anhydrides* because their water solutions are acidic. These solutions turn blue litmus red, a property of aqueous solutions of acids.

More compounds containing hydrogen are known than of any other element, even though its weight–percent abundance is not high. Hydrogen (H_2) is a colorless, odorless, and tasteless gas at ordinary temperatures. It is the lightest known substance. Although it is present in the atmosphere of the earth in only very small amounts, it appears to be a major component of the atmosphere of the sun and other stars. At ordinary temperatures, hydrogen is relatively inactive chemically, but when heated it enters into many chemical reactions. It combines with several nonmetals, forming covalent compounds, in addition to its explosive reaction with oxygen to form water. Hydrogen also reacts with some of the active metals forming ionic hydrides; in these reactions hydrogen becomes the negative ion, H^-. Hydrogen also functions as a *reducing* agent when passed over the heated oxides of metals, with the formation of the free metal and water. In removing oxygen from copper(II) oxide, for example, hydrogen acts as a *reducing* agent, and the copper(II) oxide acts as an *oxidizing* agent and furnishes oxygen for the oxidation of the hydrogen:

$$CuO + H_2 \longrightarrow Cu + H_2O$$

Reduction is defined as the gain of electrons; *oxidation* as the loss of electrons. The two processes occur simultaneously. The *reducing agent* is itself oxidized, and the *oxidizing agent* is reduced.

PROCEDURE

In this exercise you will perform either Section 1, Properties of Oxygen, or Section 2, Properties of Hydrogen—as your instructor directs. You will be informed as to the source of supply of the particular gas you will need for your tests.

1. Properties of Oxygen

A. Some Reactions of Oxygen. Four bottles of oxygen will be needed for the following tests. Litmus paper is to be used as a test for acidity or basicity (alkalinity); litmus paper is red in acidic solutions and blue in basic (alkaline) solutions.

(a) Light the end of a wood splint, blow out the flame so that only a red glow remains, and then insert the glowing end of the splint into a bottle of oxygen. Describe the result (*1*). Allow the splint to remain in the bottle until it stops burning.

Add a small amount of water to the bottle and then shake it. Add a piece of red litmus paper and a piece of blue litmus paper to the water. Describe and explain any color change in the litmus paper that you may observe (*2*).

(b) Place in a cold deflagrating spoon (see page 12, Common Laboratory Equipment) a small quantity of red phosphorus. **In the hood,** ignite the phosphorus by heating it in a flame, and lower the spoon into the second bottle of oxygen. Cover with a glass plate. Describe the result (*3*).

Add water to the bottle, shake it, and test the water with both red and blue litmus paper as in (a). Note any color change (*4*).

(c) Add enough water to the third bottle of oxygen to cover the bottom. Ignite a piece of steel wool held with tongs or in a deflagrating spoon in the Bunsen flame and quickly thrust it into the oxygen. Cover with a glass plate. Describe the result (*5*). Why cannot the litmus test be applied to the product (*6*)?

(d) Hold a 2-in. strip of magnesium ribbon with a pair of tongs. Ignite the magnesium (**Care! Keep it away from your eyes and your neighbor.**) by holding the end of the strip in the flame, and then quickly thrust the burning magnesium into the fourth bottle of oxygen. Cover with a glass plate. Describe the result (*7*).

Add water to the bottle and test the solution with litmus paper. Describe and account for the result (*8*).

B. The Thermal Decomposition of Oxygen-Containing Compounds. [*Note:* **Of the following tests, perfom the mercury (II) oxide and the lead dioxide tests in the hood.** Check with instructor first.]

Heat a 0.5-g sample of mercury(II) oxide, HgO, contained in a test tube, slowly at first, then more strongly. From time to time thrust a glowing splint into the tube. *Never allow the splint to touch the hot material*. Describe any changes observed (*9*).

Repeat the test with 0.5-g samples of manganese dioxide (MnO_2), lead dioxide (PbO_2), iron(III) oxide (Fe_2O_3), silicon dioxide (SiO_2), and calcium oxide (CaO). Record the results (*10*).

C. Catalysis. Heat 1 g of potassium chlorate ($KClO_3$) in a test tube until the salt melts and bubbles of gas are slowly evolved. Observing the necessary precautions (**eye protection**), test for oxygen with a glowing splint. Describe the result (*11*). Remove the flame.

Reminder: Goggles must be worn. Obtain a pinch of manganese dioxide (MnO_2). **Caution: Make sure the material added is manganese dioxide and not some combustible substance such as powdered charcoal, or an explosion might result.** Add the pinch of MnO_2 to the potassium chlorate and note the effect (*12*). Test for oxygen again and describe the result (*13*).

Heat the mixture until the potassium chlorate is entirely decomposed as shown by a negative test for oxygen. Allow the residue to cool, add 10 mL of water, and shake the tube until the residue has disintegrated.

Filter the resultant mixture and collect the filtrate in an evaporating dish.

Evaporate the filtrate to dryness over a flame and compare the residue to that of potassium chlorate (*14*).

Identify the residue on the filter paper (*15*). Has it undergone any apparent physical change (*16*)? Explain how this experiment demonstrates the phenomenon of catalysis (*17*).

2. Properties of Hydrogen

A. Combustion of Hydrogen. When a mixture of hydrogen and oxygen is ignited, an explosion occurs, and water is formed as the product of combustion. Because of the violence of the explosion, great caution must be used in handling hydrogen or any other combustible gas. However, *pure* hydrogen will burn quietly in air.

Obtain a test tube full of hydrogen. Holding the test tube with a test tube clamp, carry the test tube *mouth downward* to a flame **located some distance away from the generator (source). Using caution,** ignite the hydrogen in the test tube. If the gas in the test tube explodes when ignited by the flame, a mixture of hydrogen and oxygen (air) is indicated.

When a sample of pure hydrogen has been obtained, holding the test tube *mouth downward,* insert a burning splint into it. Does hydrogen support combustion (*18*)? Does the hydrogen itself burn (*19*)?

B. Density of Hydrogen. Hold an open test tube of hydrogen mouth upward for a minute, and then test with a flame. Describe and account for the result (*20*).

C. Hydrogen as a Reducing Agent. (Optional: See Instructor) Fit a combustion tube with two stoppers carrying glass tubing, as shown in Fig. 33.1. One tube is bent at right angles and is drawn to a tip of 1-mm diameter at the upper end. (For

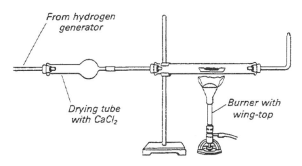

Figure 33.1 Laboratory setup for demonstrating the reducing property of hydrogen.

information on glass bending or preparing a capillary tip, see Part 3C, page 388.) Clamp the combustion tube in place and attach it to the hydrogen generator provided by means of rubber tubing. (**If a hydrogen generator as described in the Appendix is used, observe the described precautions; namely, no flames within 3 feet of the wrapped generator.** A shield of acrylic plastic is highly recommended.)

Place about 1 g of iron(III) oxide, Fe_2O_3, or copper(II) oxide, CuO, in a porcelain boat, and push the boat into the middle of the combustion tube. The system must not be a closed system.

Air in the system must first be displaced. Collect a sample of the gas by holding an inverted test tube over the exit of the combustion tube for about 30 seconds. Test the sample by carrying the inverted test tube to a flame **at least 3 feet away.** Repeat the collection of samples until the gas in the test tube burns quietly.

When the hydrogen issuing from the jet is pure, light the jet **only with a test tube of burning hydrogen.** Begin heating the combustion tube directly under the boat by means of a burner equipped with a wing top; hold the burner in your hand. A change will occur in the appearance of the solid in the boat. Describe this change (*21*). The equations for these chemical changes are

$$Fe_2O_3 + 3H_2 \longrightarrow 2Fe + 3H_2O$$

and

$$CuO + H_2 \longrightarrow Cu + H_2O$$

REPORT EXERCISE **33**

NAME ____________________ SECTION ________ DATE ________

Complete the section assigned by your instructor.

1. Properties of Oxygen

A. Some reactions of oxygen

(1)

(2)

(3)

(4)

(5)

(6)

(7)

(8)

B. The thermal decomposition of oxygen-containing compounds

(9)

(10)

C. Catalysis

(11)

(12)

(13)

(14)

(15)

(16)

(17)

2. Properties of Hydrogen

A. COMBUSTION OF HYDROGEN

(18)

(19)

B. DENSITY OF HYDROGEN

(20)

C. HYDROGEN AS A REDUCING AGENT

(21)

NAME _____ SECTION _____ DATE _____

Answer the following questions before beginning the exercise (These questions pertain to Section 1, Properties of Oxygen.):

1. Complete and balance the following equations, showing the oxides formed.

   (a) $C + O_2 \longrightarrow$

   (b) $P + O_2 \longrightarrow$

   (c) Fe (steel wool) $+ O_2 \longrightarrow$

   (d) $Mg + O_2 \longrightarrow$

2. Write equations for the combination of water with each of the above *oxides*, and classify the product as an *acidic anhydride* or *basic anhydride* (see Discussion).

   (a)

   (b)

   (c)

   (d)

3. What general classification of oxides can be made on the basis of the answers to Question 2?

4. Why do elements tested burn more brightly in oxygen than in air?

5. Complete and balance the following equations for those reactions that will take place under laboratory conditions, namely, the heat of a Bunsen flame. Consult your textbook where necessary.

   (a) $HgO \xrightarrow{\Delta}$

   (b) $MnO_2 \xrightarrow{\Delta}$

   (c) $PbO_2 \xrightarrow{\Delta}$

(d) $Fe_2O_3 \xrightarrow{\Delta}$

(e) $SiO_2 \xrightarrow{\Delta}$

(f) $CaO \xrightarrow{\Delta}$

6. What relationship exists between the activities of the metallic elements and the stabilities of their oxides?

7. Define the term *catalyst*.

34

Determination of Ascorbic Acid (Vitamin C)

OBJECTIVE

To carry out a redox titration for the determination of ascorbic acid (vitamin C), and to develop an appreciation for the role vitamins play in human biochemistry by studying the particular example of vitamin C.

DISCUSSION

A lack of vitamin C (ascorbic acid) in the diet causes the deficiency disease known as scurvy. Scurvy is characterized by skin lesions and blood vessel fragility. The structural protein *collagen* is necessary for the strength of connective tissues. When collagen is biosynthesized, it contains many residues of the amino acid *proline*. Enough of these proline residues must be converted to *hydroxyproline* residues by the enzyme *protocollagen hydroxylase* in order for the collagen to form strong fibers. This enzyme requires vitamin C, a reducing agent, as a *coenzyme* for this biological activity. The recommended daily allowance (RDA) of vitamin C for adults is about 50 mg. This RDA has specific reference only to the prevention of the dietary deficiency disease scurvy.

Vitamin C (ascorbic acid) and its derivative, dehydroascorbic acid, in addition to the above-mentioned role as coenzyme in proline hydroxylation, take part in other biological oxidation and reduction reactions—the loss and gain of hydrogen, respectively.

Vitamin C may be effective in combating the viral infection known as the "common cold." The mechanism may involve the influence of vitamin C on the synthesis and activity of *interferon*, a protective protein produced by cells infected by a virus. The RDA for this aspect of vitamin C is not established, but may range from 90–1000 mg or more, presumably influenced by human biochemical individuality.

Vitamin C is a colorless, water-soluble, crystalline solid, present in many foods, especially citrus fruits, rose hips, and many vegetables including green peppers and broccoli. Vitamin C is unstable in neutral or alkaline solution or at elevated temperatures.

The analytical method used here takes advantage of the fact that vitamin C is a strong reducing agent. The method is a redox titration; redox titrations are based on oxidation–reduction reactions. Vitamin C is the reducing agent and is oxidized; the dye 2,6-dichloroindophenol (pink in acid; blue in alkali) is the oxidizing agent and is reduced. Ascorbic acid reduces the colored dye to the colorless reduced form. The end point of the titration, carried out in acid solution, is detected by the appearance of the pink color of the excess 2,6-dichloroindophenol. Thus the dye serves not only as the oxidizing agent, but also as the indicator. The reaction is shown as follows:

Ascorbic acid (Vitamin C)   2,6-Dichloroindophenol dye (pink in acid; blue in alkali)   Dehydroascorbic acid   Reduced dye (colorless)

PROCEDURE

In this exercise you will determine the amount of vitamin C (ascorbic acid) in an unknown solution. Upon satisfactory completion of this part of the exercise you will proceed to the determination of the vitamin C content of a fruit juice that you will have provided. You should select a juice without a deep color, which would interfere with the end-point determination. On the basis of the vitamin C content of your fruit juice, you will determine the amount of this juice needed daily to achieve the RDA to prevent scurvy, assuming this juice to be the sole source of your vitamin C. And, finally, you will determine the effect of heat and exposure to air on the vitamin C content of this juice. Review the technique of titration as given in Exercise 24, and the principles of redox titrations as discussed in Exercise 28.

1. Standardization of the 2,6-Dichloroindophenol Solution

Pipet 10 mL of standard ascorbic acid solution into a 250-mL Erlenmeyer flask. Record the concentration of the standard solution, given in mg/mL, from the label (1).

Using a graduated cylinder, add 1.5 mL of glacial acetic acid (Caution! Do not inhale vapors, and avoid skin contact.) Fill a buret with 2,6-dichloroindophenol (dye) solution. Record the ini-

tial volume to 0.01 mL (2). The meniscus may be difficult to see because of the color of the dye.

Titrate the ascorbic acid solution with the dye to a faint permanent pink color. (The color may fade after the end point, but much less rapidly than during the titration.) Record the final volume to 0.01 mL (3). Repeat. If your values differ by more than 3%, repeat the titration until consistent values are obtained.

Using the average value of milliters of dye added (4), calculate the milligrams of ascorbic acid corresponding to each milliliter of dye (5). Show your calculation on Report form. Sample data: 10 mL of standard ascorbic acid (0.10 mg/mL) required 9.7 mL of dye; therefore,

$$\frac{10 \text{ mL ascorbic acid} \times 0.10 \text{ mg/mL}}{9.7 \text{ mL dye}}$$
$$= 0.10 \text{ mg ascorbic acid/mL dye}$$

2. Titration of an Unknown Ascorbic Acid Solution

Obtain an ascorbic acid solution of unknown concentration. Record the number of the unknown (6).

Pipet 10 mL of the unknown into a clean 250-mL Erlenmeyer flask. Again, with caution, add 1.5 mL of glacial acetic acid. Fill the buret with the standardized dye solution and record the intitial volume (7).

Titrate the unknown solution with the dye in the same manner as before. Record the final volume of dye used (8). Repeat until consistent values are obtained.

Using the average value of milliliters of dye used (9), and the standardized value found in (5), calculate the concentration of your unknown ascorbic acid solution in mg/mL (10). Show calculation on Report form. Sample data: 10 mL of unknown ascorbic acid required 4.8 mL of dye; therefore,

$$4.8 \text{ mL dye} \times \frac{0.10 \text{ mg ascorbic acid}}{\text{mL dye}}$$
$$\times \frac{1}{10 \text{ mL solution}}$$
$$= 0.048 \text{ mg ascorbic acid/mL}$$

3. Titration of Ascorbic Acid in Fruit Juice

Record information about kind of juice used: name, brand, size of container, vitamin C content (if provided on label), and so on (11). Strain or centrifuge juice to remove solids if necessary; assume no volume loss.

Pipet 10 mL of fruit juice into a 250-mL Erlenmeyer flask, and add 1.5 mL of glacial acetic acid. (**Caution!**) Record the initial volume of the dye in the buret (12), and titrate to the same end point as before.

Record the final buret volume (13). Repeat the procedure. Using the average value of milliliters of dye added (14), calculate the concentration of ascorbic acid in the fruit juice in mg/mL (15). Show calculation on Report form. How many ounces of this juice are necessary to achieve the RDA of 50 mg of vitamin C, knowing that 1 fluid ounce is equivalent to 30 mL (16)? How well do your data agree with those on tbe container (17)?

4. Titration of Ascorbic Acid in Heated Fruit Juice

Repeat the procedure as outlined in Section 3, with the following exception: Pipet 10 mL of fruit juice that has been previously heated on a steam bath for 10 minutes and then cooled to room temperature. Record the initial buret reading (18), final buret reading (19), and calculate the concentration of ascorbic acid in the heated fruit juice in mg/mL (20). Account for any difference between the results of Sections 3 and 4 (21).

5. (Optional) Titration of Ascorbic Acid in Fruit Juice Exposed to Air

Repeat the procedure as outlined in Section 3, with the following exception: Pipet 10 mL of fruit juice that has been exposed to the air for at least 24 hours. Record the initial buret reading (22), final buret reading (23), and calculate the concentration of ascorbic acid in the exposed fruit juice in mg/mL (24). What conclusion can be drawn about the nutritional value of this fruit juice in the diet (25)?

NAME _____ SECTION _____ DATE _____

1. Standardization of the 2,6-Dichloroindophenol Solution

(1) Concentration of standard acid, mg/mL .... _____

	Trial 1	Trial 2
(3) Final buret reading	_____ mL	_____ mL
(2) Initial buret reading	_____ mL	_____ mL
Volume of dye used (3) − (2)	_____ mL	_____ mL

(4) Average volume of dye used .... _____ mL

(5) $\dfrac{\text{mg ascorbic acid}}{\text{mL dye}}$ (show calculation) .... _____

2. Titration of an Unknown Ascorbic Acid Solution

(6) Unknown number .... _____

	Trial 1	Trial 2
(8) Final buret reading	_____ mL	_____ mL
(7) Initial buret reading	_____ mL	_____ mL
Volume of dye used (8) − (7)	_____ mL	_____ mL

(9) Average volume of dye used .... _____ mL

(10) Ascorbic acid (show calculation) .... _____ mg/mL

3. Titration of Ascorbic Acid in Fruit Juice

(11) Information about fruit juice:

	Trial 1	Trial 2
(13) Final buret reading	_____ mL	_____ mL
(12) Initial buret reading	_____ mL	_____ mL
Volume of dye used (13) − (12)	_____ mL	_____ mL

(14) Average volume of dye used .... _____ mL

(15) Ascorbic acid (show calculation) .... _____ mg/mL

(16) RDA (show calculation) .... _____ ·mg

(17)

4. Titration of Ascorbic Acid in Heated Fruit Juice

(19) Final buret reading .... _____ mL
(18) Initial buret reading .... _____ mL
       Volume of dye used (19) − (18) .... _____ mL
(20) Ascorbic acid (show calculation) .... _____ mg/mL

(21)

5. (Optional) Titration of Ascorbic Acid in Fruit Juice Exposed to Air

(23) Final buret reading .... _____ mL
(22) Initial buret reading .... _____ mL
       Volume of dye used (23) − (22) .... _____ mL
(24) Ascorbic acid (show calculation) .... _____ mg/mL

(25)

NAME _____ SECTION _____ DATE _____

Answer the following questions before beginning the exercise:

1. Vitamin C is an organic acid, and a strong reducing agent. Categorize all the other known vitamins as to type of molecule and chemical activity.

2. What common basis do vitamins have that permits them to be placed in one class of biological molecules?

3. Define the terms *enzyme, coenzyme, cofactor*, and *prosthetic group*. It will be helpful to consult a biochemistry textbook in order to answer this question.

4. List the water-soluble and the non-water-soluble vitamins. Correlate this property with human toxicity levels of the vitamins.

5. Are all of the vitamin assays the same? Explain your answer.

6. How would the experimental procedure used in this exercise have to be altered in order to determine the vitamin C content in solids such as green peppers?

7. What precautions must be taken in storing and serving foods in order to conserve their vitamin C content?

8. How many grams of dye will be reduced by 0.10 mol of vitamin C? Refer to the equation for the reaction given in the Discussion.

9. Write the structural formulas for the amino acids proline and hydroxyproline. How do they differ?

35

Classes of Organic Compounds

OBJECTIVE

To acquire knowledge of some important characteristics of organic compounds and to apply this knowledge to detecting some organic functional groups.

DISCUSSION

Organic compounds, in the modern sense, are the compounds of carbon. So numerous and important are these compounds that an entire branch of chemistry, organic chemistry, is devoted to their study. The existence of so many organic compounds is due primarily to the ability of carbon atoms to combine with other carbon atoms, forming chains of different lengths and rings of different sizes. The chemistry of organic compounds is organized by "functional groups." Compounds having the same group have similar chemical properties; thus the study of organic compounds is greatly simplified.

Carbon has four valence electrons, and shares its electrons in combination with other atoms. This bond type is the covalent bond.

The simplest organic compounds are the hydrocarbons, which contain only carbon and hydrogen atoms. Their variety is tremendous, also. Hydrocarbons can be *saturated* straight-chain molecules with increasing numbers of carbon atoms, for example:

and so on, in which all the electrons of the carbon atoms are shared with hydrogen. Hydrocarbons can also be *unsaturated* forms, in which the carbon atoms share their electrons with each other:

and so on. In each of these structure formulas that represent these molecules, the single line (—) between atoms signifies a single bond, or one pair of shared electrons, whereas the double line (==) signifies a double bond, or two pairs of shared electrons. Triply bonded molecules also exist, for example, ethyne (acetylene) $HC \equiv CH$.

Unsaturation can be determined by chemical tests, as exemplified in the first part of this exercise. The permanganate (Baeyer's) test may be summarized by the following equation:

$$\underset{\text{Purple}}{\overset{}{\ce{>C=C<}}} + MnO_4^- \longrightarrow \underset{\underset{\text{OH OH}}{}}{-\overset{|}{\underset{|}{C}}-\overset{|}{\underset{|}{C}}-} + \underset{\text{Brown}}{MnO_2} + \text{other oxidation products}$$

The bromine test may be summarized by this equation:

$$\underset{\text{Reddish-brown}}{\overset{}{\ce{>C=C<}} + Br_2} \longrightarrow \underset{\text{Colorless}}{Br-\overset{|}{\underset{|}{C}}-\overset{|}{\underset{|}{C}}-Br}$$

In addition to these straight-chain series, cyclic hydrocarbons, both saturated and unsaturated, exist. Both these straight-chain and cyclic hydrocarbons and their derivatives are classified as *aliphatic* compounds. A special homologous series beginning with benzene, and possessing unique properties, comprises the *aromatic* compounds.

Derivatives of hydrocarbons are formed by replacing one or more hydrogen atoms by groups referred to as *functional groups*. An *alcohol* is closely related to a hydrocarbon in that an alcohol group (—OH) has replaced a hydrogen atom; an alcohol is also related to water in that a hydrocarbon group has replaced a hydrogen. Thus the empirical type of formula for an alcohol is R—OH, where "R" refers to the hydrocarbon portion of the molecule aside from the functional group.

In an *organic acid*, R—COOH, the functional group is the carboxylic acid group (—COOH). This group can dissociate producing hydrogen ions, thus giving acidic properties to the hydrocarbon derivative:

$$R-COOH \rightleftharpoons R-COO^- + H^+$$

An *organic base* is an organic molecule that can accept a proton, for example, an amine, R—NH₂, with the functional amino group (—NH₂):

$$R-NH_2 + H^+ \rightleftharpoons R-NH_3^+$$

An *organic halide* is represented by the formula type R—X, where — X may be Cl, Br, or I.

Aldehydes, with the functional aldehydic group $\left(\overset{H}{\underset{}{\searrow}} C=O \right)$ and *ketones*, with

the functional ketonic group $\left(\ce{>C=O} \right)$, represent further stages of oxidation of

hydrocarbons beyond the alcohols, and behave chemically as such. The $\ce{>C=O}$ group is referred to as the *carbonyl* group.

PROCEDURE

Students may work in pairs in carrying out the tests for five of the kinds of functional groups found in organic compounds.

1. Hydrocarbons

Obtain 5-mL samples of the hydrocarbons octane (C_8H_{18}) and cyclohexene (C_6H_{10}). **(Caution: Hydrocarbons are flammable.)** Divide each sample into two portions. Label each of your four test tubes. Perform the following tests for unsaturation of these compounds.

To one portion of each of the hydrocarbons add 3 drops of slightly acidified 0.1 M potassium permanganate solution. Shake the mixture and note and record the behavior of each of the hydrocarbons (1).

Perform this test in the hood. To the remaining portion of each hydrocarbon add 5 drops of fresh bromine water (**Caution: Do not inhale or permit contact with the skin**) and shake the mixtures. Note and record the results (2).

For what functional group do these reagents test (see Discussion) (3)?

2. Alcohols

Obtain 5-mL samples of the representative alcohols ethanol (C_2H_5OH), 1-butanol (C_4H_9OH), and benzyl:

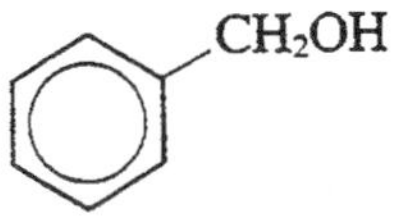

Note their color, odor, solubility in water, and action with sodium (4). **Using forceps** add a tiny piece of sodium (obtain from your instructor) to each alcohol and note the results (5). What are the products of the reaction (6)? To what inorganic reaction is this reaction related (7)? With these samples, for what functional group is this a test (8)? Write a typical equation for the sodium reaction (9).

Add to separate test tubes 1 mL of each of the alcohols: n-butyl, $CH_3(CH_2)_3OH$; sec-butyl, $CH_3CH_2(CH_3)CHOH$; tert-butyl, $(CH_3)_3COH$; benzyl, $C_6H_5CH_2OH$; and isopentyl $(CH_3)_2CHCH_2CH_2OH$. To each add 10 mL of Lucas reagent ($ZnCl_2$ dissolved in concentrated HCl). Alcohols dissolve in this reagent, but the alkyl chlorides are insoluble. Note and record the results and the time required for the reaction in each case (10). Write equations for the reactions (11). Judging from your results, how is the structure of the alcohol related to the reaction rate (12)?

3. Acids and Bases

Obtain small samples of benzoic acid, C_6H_5COOH; acetic acid, CH_3COOH; and naph-

thylamine, $C_{10}H_7NH_2$. (**Caution: Do not inhale or permit contact with the skin.**) Test each of these for solubility in water, 0.1 M hydrochloric acid, and 0.1 M sodium hydroxide, by adding a few crystals (or a few drops if a liquid) of the organic substances to 3-mL portions of each of the solvents named (13). Substances soluble in water can be tested with litmus paper (14).

When is the litmus-paper test applicable for organic acids and bases (15)? What tests can be made in cases in which the litmus test is not applicable (16)? What general statement can be made concerning the solubility of salts of organic acids and bases in organic solvents and in water (17)?

4. Organic Halides

Add a few drops of 0.1 M silver nitrate solution first to 2 mL of 0.1 M sodium bromide solution and then to 2 mL of n-butyl bromide, C_4H_9Br. Shake the mixtures vigorously. Describe and account for the results (18). Write equations for any reactions (19).

Make a spiral by winding a 3-in. length of bright; copper wire around a pencil. Hold the wire in a colorless Bunsen flame. Dip the wire into 1 or 2 drops of n-butyl bromide on a watch glass and hold the wire in the edge of the flame. Note the color produced (20). Also test dichloromethane (21). This is a sensitive test for halogen.

5. Aldehydes

Fehling's solution, essentially a solution containing copper(II) ions as tartrate complexes, is an indicator of the presence of reducing groups such as the aldehyde group. The reaction type is the following:

$$RCHO + 2Cu(OH)_2 \longrightarrow$$
$$RCOOH + Cu_2O \downarrow + 2H_2O$$

Reddish-brown
color

Place 10 mL of a mixture of equal volumes of Fehling's solutions A and B in a test tube and add 5 to 10 drops of a 5% formaldehyde solution. (**Caution: Do not inhale.**) Boil the mixture gently for 2 or 3 minutes and observe and record the results (22).

Repeat the test with a ketone such as acetone (23).

Can aldehydes be differentiated from ketones by this test (24)?

Repeat the Fehling's test with several representative sugars such as sucrose, glucose, maltose, lactose, and fructose, using a few drops of a 5% solution of each sugar for each test. Record your results (25).

Which of these sugars appear to have reducing groups (26)?

NAME _____ SECTION _____ DATE _____

1. Hydrocarbons

(1), (2) Describe results:

Hydrocarbon	Acidified permanganate	Bromine-H_2O
Octane		
Cyclohexene		

(3)

2. Alcohols

(4), (5) Describe:

Alcohol	Color	Odor	Solubility in H_2O	Action of Na
Ethanol				
1-Butanol				
Benzyl alcohol				

(6)

(7)

(8)

(9) Equation:

(10), (11) Results of Lucas test:

Alcohol*	Results	Time	Equation for reaction
n-Butyl (1°)			
sec-Butyl (2°)			
tert-Butyl (3°)			
Benzyl (1°)			
Isopentyl (1°)			

*1° = primary alcohol; 2° = secondary alcohol; 3° = tertiary alcohol: A classification scheme depending on where the −OH group attaches to the hydrocarbon chain.

(12)

3. Acids and Bases

(13), (14) Indicate solubility as s, is, or sl s:

Substance tested	Medium			Results of litmus test
	H_2O	HCl	NaOH	
Benzoic acid				
Acetic acid				
Naphthylamine				

(15)

(16)

(17)

NAME _____ SECTION _____ DATE _____

4. Organic Halides

(18)

Substance tested	Description of AgNO₃ test	Reason for result
Sodium bromide		
n-Butyl bromide		

(19) Equation(s):

(20) Color: _____

(21) Color: _____

5. Aldehydes

(22)

(23)

(24)

(25), (26) Check if test is positive; check if reducing groups are present.

Sugar	Fehling's test	Reducing groups
Sucrose		
Glucose		
Maltose		
Lactose		
Fructose		

NAME _____ SECTION _____ DATE _____

Answer the following questions before beginning the exercise:

1. Underline the following hydrocarbons which are *unsaturated:*
   (a) CH_4
   (b) $CH_3CH{=}CH_2$
   (c) $CH_3C{\equiv}CH$

2. The condensed structural formula for *n*-butyl alcohol, C_4H_9OH,
   is $CH_3CH_2CH_2CH_2OH$ (*n*- means straight chain).
   The corresponding formula for *sec*-butyl alcohol, $CH_3CH_2(CH_3)CHOH$,
   which has a branched chain, is

$$CH_3CH_2{-}\overset{\displaystyle H}{\underset{\displaystyle CH_3}{C}}{-}OH$$

   Write out, in a similar manner, the formula for *tert*-butyl alcohol,
   $(CH_3)_3COH$:

3. Classify each of the following compounds from its functional group (see
   Discussion):
   (a) CH_3COOH _____   (d) $CH_3CH_2CH_3$ _____

   (b) 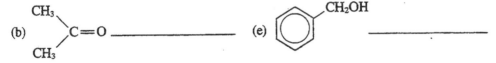 _____   (e) _____

   (c) 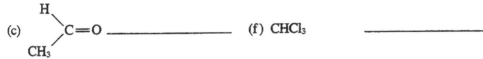 _____   (f) $CHCl_3$ _____

4. In organic chemistry, what is the value of studying the chemical behavior of
   functional groups?

The Covalent Bond: Geometry, Isomerism, Conformation

OBJECTIVE

To comprehend the geometric relations of atoms in some simple covalently bonded molecules and the theoretical basis accounting for these relations.

DISCUSSION

In recent years, data obtained from spectrometric measurements, x-ray and electron diffraction experiments, and other sources have yielded precise information about bond distances, angles, and energies. In many cases, the data confirmed conclusions reached earlier. In other cases, valuable new insights were acquired. Structure theory has advanced far beyond the electron-dot representations and now rests securely on the foundations of quantum and wave mechanics. Although problems involving only simple molecules can now be solved with mathematical rigor, such approximations as the valence bond theory and the molecular orbital theory have been extremely successful in securing results that agree with experimental measurements. Several kinds of structural models are available commercially. Particularly suitable for student use, because of low cost and simplicity, are the ball-and-stick sets and the framework sets. Your instructor will probably supplement these with accurately scaled, space-filling models, or plastic Minit model kits.

PROCEDURE

You will probably perform this exercise in groups, designated by your instructor.

1. Geometry

A. Tetrahedral Configuration. Construct models for methane, CH_4, in two ways: one with tetrahedral angles and the other with the hydrogen atoms at the corners of a square and the carbon atom at the center. For the first way, use a ball-and-stick model set; for the other, use a framework model set. If a framework model set is not available, use a graphing method: Draw a square of suitable, workable dimensions, place hydrogen atoms at the corners and a carbon atom in the center, as shown in Fig. 36.1.

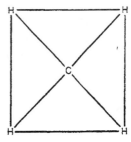

Figure 36.1 A graphical representation of a square-planar methane model.

Substitute a chlorine atom for a hydrogen atom in each model. How many isomers are obtainable (*1*)? Would this be true for other arrangements, for example, one with some of the angles between hydrogen atoms smaller than others (*2*)? Since only one chloromethane is known, what conclusion follows (*3*)?

Now substitute a second chlorine atom for a hydrogen. How many isomers are obtainable from the tetrahedral and the square-planar models (*4*)? Only one dichloromethane exists. What does this indicate about the correctness of the tetrahedral and square-planar configurations (*5*)?

Electron diffraction experiments show that there are two interatomic distances in the ratio of 1.64 in carbon tetrachloride, CCl_4. These are most likely the C—Cl (read carbon to chlorine) and the Cl—Cl distances. Construct a tetrahedral and a square-planar model for this compound with the framework set or graphically, making the bond lengths correspond to the known lengths. Measure the Cl—Cl distance for each model (*6*). Compute the ratio of the two distances for each model (*7*). Which model is correct (*8*)?

B. Water and Ammonia. Construct models of H_2O and NH_3 on the assumption that the bonds are sp^3 hybrids. Draw models of these molecules on the assumption that the bonds involve only the p orbitals of oxygen and nitrogen (*9*). The bond angles have been found to be 104.5° and 106.8°, respectively; how can these observed angles be explained (*10*)? When NH_3 is protonated, what is the configuration of the resulting ion (*11*)?

C. A Lewis Acid–Base Reaction. According to the Lewis concept, an acid is any species (molecule or ion) that can accept a pair of electrons; a base is any species (molecule or ion) that can donate a pair of electrons.

Make a model of the BF_3 molecule, remembering that there are only five electrons in the boron atom. Show how BF_3 combines with NH_3 to form F_3B—NH_3. Specify all angles before and after combination (*12*). Show changes in bond type (*13*).

2. Isomerism

A. Single Bonds. Construct a model of ethane, CH_3CH_3, from either the ball-and-stick set or the framework set. Substitute two chlorine atoms for

two hydrogen atoms. How many isomers are possible (*14*)? Only one compound having the two chlorine atoms on different carbon atoms is known. What property of the C—C bond is indicated (*15*)?

B. Multiple Bonds. Construct a ball-and-stick model for ethene CH_2＝CH_2, using the springs to represent the double bond. Substitute a chlorine atom for a hydrogen on each carbon. There are two compounds known that can be designated 1,2-dichloroethene. What does this fact indicate about the C—C double bond (*16*)?

Explain this conclusion in terms of sp^3 and sp^2-π bond types (*17*). What are the angles predicted between the C—H bonds inferred from these two bond types (*18*)?

The measured angle in ethene itself is 116.7°. What does this suggest (*19*)?

Draw the structure for ethyne (acetylene),

$$CH \equiv CH$$

What bond types are present (*20*)?

3. Conformation

A. Straight-Chain Compounds. The different arrangements of atoms that can be converted into one another by rotation about single bonds are called *conformations*. Figure 36.2 shows ways of representing conformations by diagrams, using ethane as an example: Fig. 36.2(a) shows the "saw-horse" method of representing the eclipsed conformation, Fig. 36.2(b), the Newman projection method of representation. The sawhorse representation of the staggered conformation of ethane is shown in Fig. 36.2(c); the Newman projection is shown in Fig. 36.2(d).

Take your model of ethane; view it down the C—C bond and twist the carbon atoms until you reproduce first the staggered, and then the eclipsed conformation. Staggered conformations have lower torsional energies and hence are more stable than eclipsed conformations. Draw the conformations of lowest energy (greatest stability) (*21*). Either Newman projections or sawhorse drawings are suitable for this.

B. Carbon Ring Systems. Construct models for cyclopropane, cyclobutane, cyclopentane, and cyclohexane from the framework set, assuming sp^3

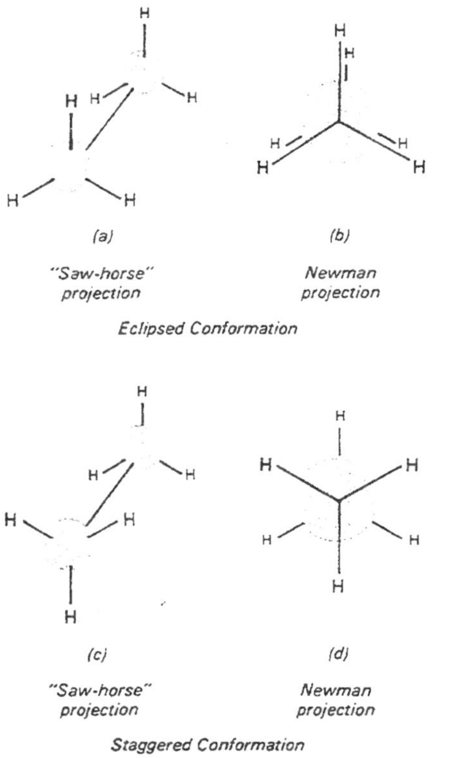

Figure 36.2 Conformations of ethane; (a) "sawhorse" projection and (b) Newman projection of the eclipsed conformation; and (c) "sawhorse" projection and (d) Newman projection of the staggered conformation.

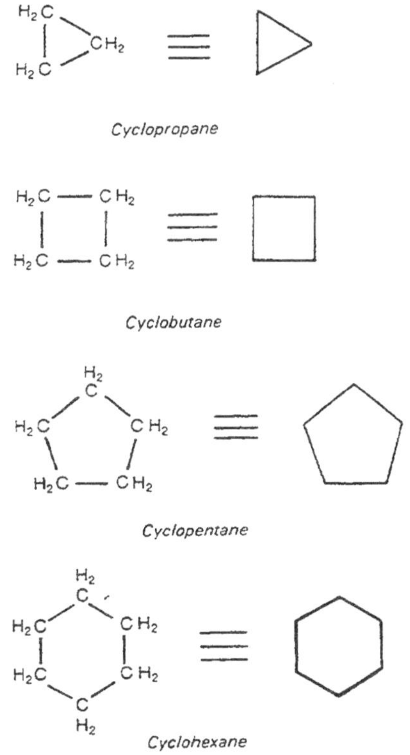

Figure 36.3 Formulas for cyclopropane, cyclobutane, cyclopentane, and cyclohexane.

hybridization. The formulas for these compounds are given in Fig. 36.3. Which of these substances has the most nearly normal bond angles (22)? Which of these substances has the least stable (highest energy) C—C bonds (23)?

Why are the molecules of cyclohexane nonplanar (24)? Manipulate the model of cyclohexane by twisting; how many conformations result (25)? Are rings of more than six carbon atoms likely to be stable? Explain (26).

Remove one of the "hydrogens" and replace it with "chlorine." How many conformations are possible for monochlorocyclohexane by twisting the model (27)? Are they equally probable (28)? If chlorine were replaced by a large group, such as $(CH_3)_3C$—, would one of the conformations be disfavored (29)?

REPORT EXERCISE **36**

NAME ______________________ SECTION ________ DATE ________

1. Geometry

A. Tetrahedral configuration

(1)

(2)

(3)

(4)

(5)

(6)

(7)

(8)

B. Water and ammonia

(9)

(10)

(11)

C. A Lewis acid–base reaction

(12)

(13)

2. Isomerism

A. Single bonds

(14)

(15)

B. Multiple bonds

(16)

(17)

(18)

(19)

(20)

NAME ______________________ SECTION ________ DATE __________

3. Conformation

A. Straight-chain compounds

(21)

B. Carbon ring systems

(22)

(23)

(24)

(25)

(26)

(27)

(28)

(29)

NAME _____ SECTION _____ DATE _____

Answer the following questions before beginning the exercise:

1. By means of atomic orbital diagrams, show
   (a) sp^3 hybridization of the carbon atom:

   (b) sp^2 hybridization of the carbon atom:

   (c) sp hybridization of the carbon atom:

2. Referring to ethene as an example, explain what is meant by a π bond:

3. (a) Two isomers of dichlorinated ethene are shown below:

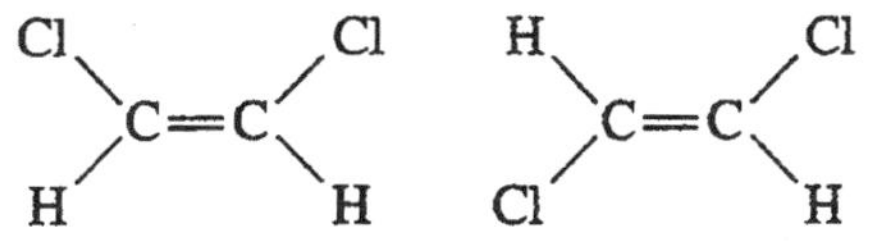

   Can these be separated? If so, suggest a method. (*Note:* They behave the same chemically.)

   (b) Two conformers of ethane are shown in Fig. 36.2—eclipsed and staggered. Can these be separated? Explain.

313

Organic Compounds of Commercial Significance

OBJECTIVE

To become familiar with several commercially important organic compounds and with the technique and principle of esterification.

DISCUSSION

The organic compounds of commercial significance portrayed in this exercise, only a small fraction of the total number existing, are perfumes and flavors, dyes, plastics, and detergents. The distinctive and attractive odors and flavors of many flowers and ripe fruits are due to the presence of one or more esters. *Esters* are the products of reaction of acids with alcohols, that is,

$$R-\overset{\overset{\displaystyle O}{\|}}{C}-OH + R'OH \longrightarrow R-\overset{\overset{\displaystyle O}{\|}}{C}-OR' + H_2O$$

The formula type for an ester is

$$R-\overset{\overset{\displaystyle O}{\|}}{C}-OR'$$

The significance of the formula type and the R group was discussed in Exercise 35. The symbol "R'" indicates that the two R— groups in the ester formula need not be the same.

Although the term *detergent* means cleansing agent, and includes soap, it is now commonly used to refer to soap substitutes. *Soaps* are sodium salts of fatty acids, whereas *detergents* are sulfates or sulfonates of long-chain alcohols. The following formula for sodium stearate represents a typical soap:

$$CH_3(CH_2)_{15}CH_2COO^-Na^+$$
Sodium stearate

The following formula for sodium lauryl sulfate represents a typical detergent:

$$CH_3(CH_2)_{10}CH_2OSO_3^-Na^+$$
Sodium lauryl sulfate

The molecules of both soaps and detergents consist of long hydrocarbon chains attached to a polar group. The hydrocarbon end (hydrophobic) is attracted by the dirt or oil particles, and the polar group (hydrophilic) by the water. This allows suspension of the dirt particles in the solution as colloidal particles, and thus they are washed away.

PROCEDURE

In this exercise some examples of commercial organic substances are considered; namely, perfumes and flavors, soaps, and detergents. You may be assigned one or both of these experiments by your instructor.

1. Perfumes and Flavors

In this experiment when you are asked to "smell" or "note odor" of a substance, *waft* some of the odor toward you with your hand; do not breathe in directly or deeply.

Smell samples of cinnamic aldehyde, benzaldehyde, methyl benzoate, methyl butyrate, and vanillin. Name naturally occurring materials that have similar odors (*1*).

Place a very small amount of salicylic acid, o-HOC_6H_4COOH, in a test tube, and add one drop of 3 M sulfuric acid, 3 drops of water, and 3 drops of methanol. Stopper loosely and place in hot water for about 10 minutes. Describe the odor (*2*). Write the equation for the reaction (*3*).

Place 3 drops of pentanol, $C_5H_{11}OH$, 2 drops of glacial acetic acid, and one drop of 3 M sulfuric acid in a test tube. Place the test tube in hot water for about 10 minutes. Note the odor and identify the natural product whose odor it resembles most closely (*4*). Write the equation for the reaction (*5*).

Warm a mixture of 1 mL of ethanol, 0.2 g of *beta*-naphthol,

and 5 drops of concentrated sulfuric acid. What natural product has an odor resembling the odor of the product of this reaction (*6*)? What compound is formed (*7*)? Does this compound belong to the same class as those prepared above (*8*)? Write the equation for the reaction (*9*).

2. Detergents

A. Preparation of Soap. In this experiment use caution when working with the solid sodium hydroxide or its solution. **Do not allow the sodium hydroxide to touch the face, hands, or clothing.** If it does, wash the area immediately and thoroughly with water.

Do your weighing quickly, and keep the reagent bottle closed when not in use because sodium hydroxide readily picks up moisture from the air.

Make a solution of sodium hydroxide by gradually adding, with stirring, 3 g of sodium hydroxide pellets to 5 mL of water contained in a beaker. Stir until dissolved.

Place about 10 g of tallow, cottonseed oil, or lard in a 400-mL beaker. Immerse the beaker in hot water; then add the sodium hydroxide solution to the beaker. Heat the mixture gently in the water bath for about 40 minutes and stir it from time to time. When the product has hardened, cut it into slices and dissolve in about 200 mL of hot water. Add 100 mL of saturated sodium chloride solution and filter off the precipitated soap. Allow it to dry; examine and describe the product (*10*). Glyceryl stearate $(C_{17}H_{35}COO)_3C_3H_5$, an ester of stearic acid, $C_{17}H_{35}COOH$, and the trihydroxy alcohol glycerol, $C_3H_5(OH)_3$, is a typical component of fats and oils.

The formation and hydrolysis of glyceryl stearate is shown as follows:

$$3C_{17}H_{35}COOH + C_3H_5(OH)_3 \underset{\text{Hydrolysis}}{\overset{\text{Esterification}}{\rightleftharpoons}}$$

Stearic acid   Glycerol

$$(C_{17}H_{35}COO)_3C_3H_5 + 3H_2O$$

Glyceryl stearate.   Water

The esterification reaction, shown structurally, is as follows:

$C_{17}H_{35}COOH$ $HO-CH_2$

$C_{17}H_{35}COOH + HO-CH$ $\longrightarrow$

$C_{17}H_{35}COOH$ $HO-CH_2$

$$C_{17}H_{35}COOCH_2$$
$$C_{17}H_{35}COOCH + 3H_2O$$
$$C_{17}H_{35}COOCH_2$$

Write the equation for the *saponification* (alkaline hydrolysis) of glyceryl stearate (*11*).

B. Properties of Synthetic Detergents. Dissolve a small piece of the soap you made, or other, such as pure castile soap, in 50 mL of water. Dissolve a like amount of a synthetic detergent in 50 mL of water. The soap and synthetic detergent solutions *must* be free of solid particles. Filter if necessary. Perform the following tests:

Add 10 drops of cottonseed oil to 10 mL of water. Divide the mixture into three portions, in three separate test tubes. Use a test tube rack. Add nothing to the first; add 1 mL of soap solution to the second; and add 1 mL of synthetic detergent solution to the third. Shake all three test tubes and de-scribe the results (*12*). What type of action is exhibited here (*13*)?

For this test, you will need six test tubes, properly marked, and a test tube rack. Test 5-mL portions of the soap and synthetic detergent solutions with 1-mL portions of 0.1 M solutions of hydrochloric acid, calcium nitrate, and magnesium sulfate. Note and record the results (*14*). Which type of detergent would you expect to be better for use with hard water and which for acidic water (*15*)? Hard water contains dissolved calcium, magnesium, and iron salts (see Exercise 31).

To 1 mL of soap solution add 5 mL of water and 1 drop of phenolphthalein solution. Note and record the result (*16*). Mix 1 mL of soap solution, 4 mL of water, 4 mL of ethanol, and 1 drop of phenolphthalein solution. Compare the results with the above test (*17*). This is a test for free alkali. Repeat the test using the synthetic detergent in place of the soap (*18*).

Optional. A commercial soap (for example, Dial, or Ivory) may be tested according to the preceding three paragraphs, if your instructor so directs (*19*).

NAME ______________________ SECTION ________ DATE ________

1. Perfumes and Flavors

(1)

Sample	*Similar natural odor*
Cinnamic aldehyde	
Benzaldehyde	
Methyl benzoate	
Methyl butyrate	
Vanillin	

(2)

(3) Equation:

(4)

(5) Equation:

(6)

(7)

(8)

(9) Equation:

2. Detergents

(10)

(11) Equation:

(12)

Mixture	Effect
Oil + Soap solution	
Oil + Detergent	
Oil + Water	

(13)

(14) Describe results:

	Solution added		
Solution	HCl	$Ca(NO_3)_2$	$MgSO_4$
Soap			
Detergent			

(15)

(16)

(17)

(18)

NAME _____ _ SECTION _____ DATE _____

(19) (Optional)

NAME _____ SECTION _____ DATE _____

Answer the following questions before beginning the exercise:

1. An ester is formed from the reaction of an acid with an alcohol. With this in mind, complete the following equation:

$$CH_3COOH + CH_3CH_2OH \longrightarrow$$

2. List the two "R" groups in the above equation:

3. Write the structure of glycerol. What kind of a compound is glycerol?

4. What are the products of *hydrolysis* of an ester?

5. What are the products of *saponification* of an ester?

6. How does a synthetic detergent differ chemically from a soap?

7. Explain how a soap or detergent is effective in cleaning away dirt.

8. Why does Ivory soap float?

Preparation of an Ester: Acetylsalicylic Acid (Aspirin)

OBJECTIVE

To become familiar with the technique and principle of esterification.

DISCUSSION

Aspirin is a drug widely used as an antipyretic agent (to reduce fever), as an analgesic agent (to reduce pain), and/or as an anti-inflammatory agent (to reduce redness, heat, or swelling in tissues). Chemically, aspirin is an *ester*. Esters are the products of reaction of acids with alcohols, as shown in the following equation using type formulas:

$$\underset{\text{An acid}}{R-\overset{\overset{\displaystyle O}{\|}}{C}-OH} + \underset{\text{An alcohol}}{R'-OH} \longrightarrow \underset{\text{An ester}}{R-\overset{\overset{\displaystyle O}{\|}}{C}-OR'} + H_2O$$

The symbol R refers to the hydrocarbon portion (radical) of the molecule aside from the functional group. In an organic acid, $R-\overset{\overset{\displaystyle O}{\|}}{C}-OH$, the functional group is the carboxylic acid group $-\overset{\overset{\displaystyle O}{\|}}{C}-OH$. The type formula for an alcohol is $R-OH$, where the functional group is the hydroxyl group ($-OH$). The symbol R' indicates that the two R— groups in the ester formula need not be the same. It has been shown by radioactive tracer methods that in the mechanism of the esterification reaction, the —OH group is split from the acid and the —H from the alcohol.

Aspirin can be made as follows (refer to the above equation):

$$\underset{\text{Acetic acid}}{CH_3-\overset{\overset{\displaystyle O}{\|}}{C}-OH} + \underset{\substack{\text{Salicylic acid}\\ \text{(containing an}\\ \text{—OH group)}}}{HO-\!\!\!\bigcirc\!\!\!\overset{\overset{\displaystyle O}{\diagup\!\!C}-OH}{}} \rightleftharpoons \underset{\substack{\text{Aspirin}\\ \text{(acetylsalicylic acid,}\\ \text{an ester)}}}{CH_3-\overset{\overset{\displaystyle O}{\|}}{C}-O-\!\!\!\bigcirc\!\!\!\overset{\overset{\displaystyle O}{\diagup\!\!C}-OH}{}} + H_2O$$

The use of acetic anhydride instead of acetic acid, however, is a better preparative method, because the anhydride reacting with the water to form acetic acid tends to drive the reaction to the right as shown below. An acid catalyst also is used to speed up the reaction.

| Acetic anhydride | Salicylic acid | Aspirin | Acetic acid |

The *theoretical yield* of aspirin in grams is the mass calculated from the above equation, with salicylic acid being the limiting reagent. The *actual yield* of aspirin in grams is obtained by carrying out the synthesis: item (*3*) in the Report form. The *percentage yield* is found by the following relationship:

$$\frac{\text{Actual yield}}{\text{Theoretical yield}} \times 100$$

Phenols are a class of compounds in which the —OH group is attached to an aromatic ring. Note that this group is present in salicylic acid, but should not be present in *pure* acetylsalicylic acid. Many phenols form colored complexes with ferric chloride. The colors range from green through blue and red through violet. Hence, a 1% $FeCl_3$ solution is employed as a test for the presence of phenols.

PROCEDURE

In this exercise, you will prepare a sample of aspirin, determine your theoretical, actual, and percentage yields, and estimate the purity of your product based on its melting point and a test for phenolic impurities.

Obtain a clean, dry 50-mL Erlenmeyer flask. Add 2.0 g of salicylic acid, 5.0 mL of acetic anhydride **(Caution: Avoid contact with the skin, and do not inhale),** and 5–10 drops of concentrated sulfuric acid **(Caution: Causes severe skin burns)** to the Erlenmeyer flask. Mix the contents by swirling gently until the reactants are in solution. Then heat the flask and contents by clamping it in a water bath maintained at 80°C for 15 minutes.

Now remove the flask from the water bath and immediately, but **cautiously,** add 2 mL of water to decompose any remaining excess acetic anhydride. Allow the flask to cool to room temperature. Then add 20 mL of water, stir well, and immediately place the flask in an ice-water bath. Crystals should begin to appear. Sometimes it may be necessary to scratch the inside of the flask with a glass rod to initiate crystallization.

While waiting for the crystallization, assemble a suction filtration setup (see Fig. D.4 and also Fig. D.5 in Part 3), and familiarize yourself if necessary, with the procedure. If suction filtration is not possible, substitute gravity filtration (see Fig. D.3). When crystallization seems complete, collect the aspirin on filter paper by suction filtering the mixture. Wash the aspirin with two separate 10-mL portions of ice-cold distilled water. Suction the product as dry as possible.

Carefully remove the product and filter paper from the funnel, place on a watch glass, and allow to air dry. Weigh a dry sheet of the same type of filter paper to the nearest 0.1 g (*1*). When the product is dry, weigh the product plus filter paper to the nearest 0.1 g (*2*). The difference is the mass of your product, or actual yield (*3*).

Determine the melting point of your completely dry (may take overnight) product (see Fig. K.1 in Part 3) using an oil bath; heat the oil bath slowly and report the melting point as the temperature at which the last crystals disappear (*4*). Look up the melting point of aspirin in a handbook, and record it (*5*). Compare these two melting points, and comment on the purity of your product (*6*).

Calculate your theoretical yield (*7*) and your percentage yield (*8*).

In separate test tubes, place a few crystals of:

(a) salicylic acid
(b) your aspirin preparation
(c) commercial aspirin

Dissolve each in 5 mL of methanol. Add 3 drops of 1% ferric chloride solution to each test tube, shake gently, and note and record the immediate color (9). Does your aspirin contain any unreacted salicylic acid (see Discussion) (10)? Explain (11).

NAME _____ SECTION _____ DATE _____

Weight of salicylic acid used .... ——————— g

Weight of acetic anhydride used .... ——————— g
   (density = 1.08 g/mL)

(1) Weight of filter paper .... ——————— g

(2) Weight of filter paper + aspirin .... ——————— g

(3) *Actual yield* of aspirin (2) − (1) .... ——————— g

(4) Melting point of the aspirin product .... ——————— °C

(5) Melting point of aspirin from handbook .... ——————— °C

(6) Comment on purity of product, based on a comparison of (4) and (5):

(7) *Theoretical yield* of aspirin (show calculation):

(8) *Percentage yield* of aspirin (show calculation):

(9)

Product	Color with FeCl₃ test
Salicylic acid	
Your aspirin	
Commercial aspirin	

---

(10)

(11)

NAME _____ SECTION _____ DATE _____

Answer the following questions before beginning the exercise:

1. What is the general structure (type formula) of esters?

2. How are esters prepared?

3. What is the carboxylic acid group?

4. Why is acetic anhydride, rather than acetic acid, used in the preparation of aspirin?

5. 1.02 g of acetylsalicylic acid were obtained from 1 g of salicylic acid by reaction with excess acetic anhydride. Calculate the percentage yield of acetyl-salicylic acid.

6. Methyl salicylate, another ester, can be prepared from salicylic acid and methyl alcohol. Draw the structure of this ester.

7. Describe the ferric chloride test (see Discussion).

39

Carbohydrates

OBJECTIVE

To become familiar with one of the classes of organic compounds that are essential for life.

DISCUSSION

Carbohydrates constitute one class of biologically important organic molecules (also see Exercise 40). Carbohydrates are compounds composed of carbon, hydrogen, and oxygen—the ratio of hydrogen to oxygen being $2:1$. (Derived carbohydrates may also contain other elements.) Some carbohydrates such as starch and cellulose are very large, complex molecules, called macromolecules. These macromolecules (polymers) are composed of simpler "building block" compounds (monomers) bonded together.

The basic carbohydrate molecules are simple sugars, or *monosaccharides* (polyhydroxy aldehydes or ketones). All simple sugars, when in open-chain form, contain a carbonyl

$$\overset{|}{\underset{|}{C}}=O$$

group; if the double-bonded oxygen is attached to the terminal carbon of the carbon chain, the combination is an *aldehyde* group, whereas if it is attached to a nonterminal carbon, the combination is a *ketone* group. Many different sugars exist; the carbon chain that forms the sugar can be of different lengths (3–7 carbons). Shown in Fig. 39.1 are open-chain (Fischer) formulas of D-glucose (an aldohexose) and of D-fructose (a ketohexose). The prefix D means that the absolute configuration at the asymmetric carbon furthest from the aldehyde or ketone group, namely C-5, is the same as in D-glyceraldehyde. D-glyceraldehyde has this absolute configuration.

$$
\begin{array}{c}
H \\
| \\
C=O \\
| \\
H-C-OH \\
| \\
CH_2OH
\end{array}
$$

The term *hexose* refers to the six carbon atoms in the chains. Both of these sugars have the same molecular formula, $C_6H_{12}O_6$, and are structural isomers. The predominant forms of glucose and fructose in solution are not the open-chain structures, but rather the cyclized or ring structures. Modified Fischer projections have been used to depict the cyclic intramolecular hemiacetal form. For example, α-D-glucose may be represented as follows:

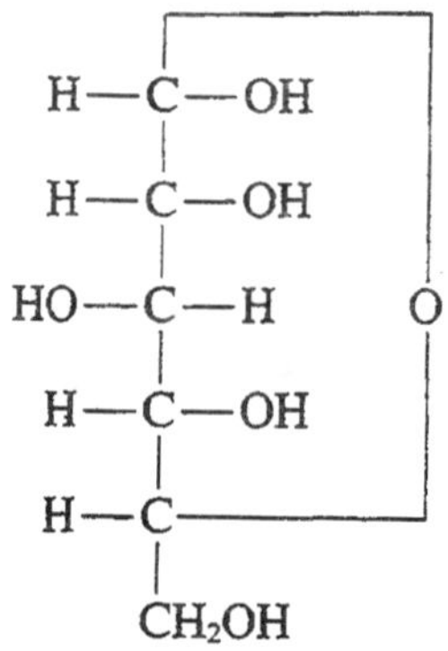

Haworth introduced an alternative projection formula, the sugar ring being written as a planar hexagon with the oxygen in the upper right vertex. Substituents are indicated by straight lines through each vertex, either above or below the plane. The ring structures are represented by Haworth projection formulas shown beneath the corresponding open-chain formulas in Fig. 39.1. The plane of the ring is perpendicular to the plane of the paper, with the heavy line on the ring closest to the reader. Conformation structures are not included. Carbohydrates known as *disaccharides* (such as lactose, maltose, and sucrose), so-called "double sugars," are composed of two monosaccharides bonded together in sequence. *Polysaccharides*, such as starch,

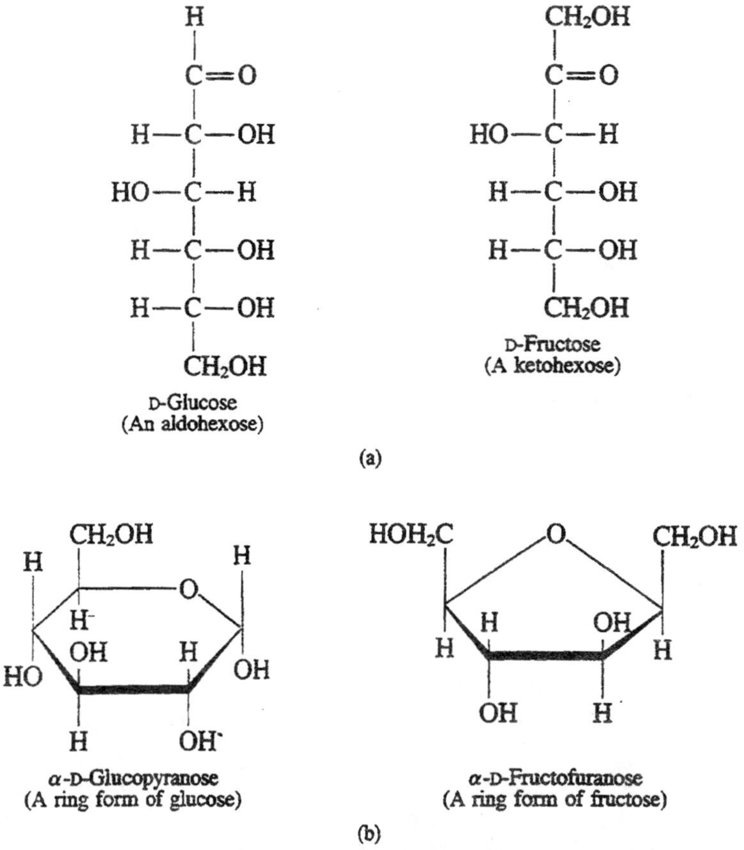

(a)

(b)

Figure 39.1 Formulas of glucose (an aldohexose) and of fructose (a ketohexose); (a) open-chain formulas (Fischer), and (b) projection ring formulas (Haworth).

are polymers containing many monosaccharide (monomer) units bonded together in long chains. Some polysaccharides exist in straight chains; others in branched chains. The biosynthesis of polysaccharides from monosaccharides takes place by a process known as dehydration synthesis; the reverse reaction is known as hydrolysis, or degradation.

Many different tests are available that allow identification of unknown carbohydrates. The tests that this exercise will involve are the following:

1. Test for Reducing Sugars: Fehling's Test. Reducing sugars are sugars, either mono- or disaccharides, which have a *free* aldehyde or ketone group available for reaction with the Cu^{2+} ion provided by the Fehling's solution:

$$Cu^{2+} + \text{reducing sugar} \longrightarrow \underset{\text{Red ppt}}{Cu_2O \downarrow} + \text{oxidized sugar}$$

Fehling's solution is a freshly mixed solution of A, cupric sulfate solution; and B, sodium hydroxide and sodium potassium tartrate solution. The tartrate ion forms a chelate complex that decreases the Cu^{2+} ion concentration below that necessary for the precipitation of cupric hydroxide.

2. Test for Polysaccharides: Iodine Test. The deep blue color obtained when applying this test to a polysaccharide such as starch is believed to be due to a coordination complex between the helically coiled polysaccharide chain and the iodine centrally located within the helix. Variation in intensity and shade of color depend on the size of the molecule and on temperature.

3. Test to Distinguish between Mono- and Disaccharide Reducing Sugars: Barfoed's Test. A differential rate of reaction with acidic cupric acetate solution is the basis of this test:

$$Cu^{2+} + \text{reducing monosaccharide} \xrightarrow{H^+} Cu_2O \downarrow + \text{oxidized monosaccharide}$$
$$\text{(Fast reaction)}$$

$$Cu^{2+} + \text{reducing disaccharide} \xrightarrow{H^+} Cu_2O \downarrow + \text{oxidized disaccharide}$$
$$\text{(Slow reaction)}$$

4. Test for Fructose: Seliwanoff's test. Concentrated HCl acts on fructose to produce a dehydration product which when condensed with resorcinol quickly yields a deep red color, a positive test for a ketohexose such as fructose.

5. Test for Lactose: Mucic Acid Test. Lactose, when treated with concentrated HNO_3, is oxidized to mucic acid, the only saccharic acid insoluble in cold water.

PROCEDURE

In this exercise you will be given an unknown carbohydrate that may be one of the following: starch, lactose, maltose, sucrose, fructose, or glucose. As a means of identifying your unknown you will use the series of reactions indicated in the accompanying flow diagram (p. 336). You need to perform only those tests necessary for a positive identification of your unknown. However, for the sake of experience, it would be beneficial for you to become acquainted with all of the listed tests. Along with your unknown you will be running some known samples in order for you to be able to interpret the test results. Record your data, as you obtain it, in the Report form.

Prepare 20 mL of a 1% solution of your unknown. Record its number in the Report form.

FLOW DIAGRAM: POSSIBLE UNKNOWN FRUCTOSE,
GLUCOSE, MALTOSE, STARCH, OR SUCROSE

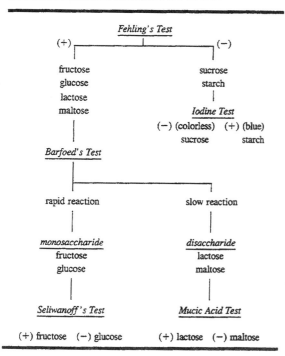

1. Fehling's Test

Mix 15 mL of Fehling's solutions A and B; use the freshly mixed solution in the following tests. Add 1 mL of 1% solutions of fructose, glucose, sucrose, and your unknown to four separate labeled test tubes. To each test tube add 5 mL of the mixed Fehling's solution. Place the test tubes in a boiling water bath and observe the color produced in each. A positive test is a red precipitate of Cu_2O.

If this test is positive, proceed to Barfoed's test; if this test is negative, proceed to the iodine test.

2. Iodine Test

Place 5 mL of 1% solutions of glucose, starch, your unknown, and 5 mL of distilled water into four separate labeled test tubes. Add 3 drops of dilute iodine solution to each test tube and compare the color produced with that obtained with the distilled water blank.

3. Barfoed's Test

Place 5 mL of 1% solutions of glucose, lactose, and your unknown into three separate labeled test tubes. To each test tube then add 5 mL of Barfoed's reagent and heat in a water bath for 10 minutes. Compare the rate of the formation of the red precipitate Cu_2O in your unknown with that in the two other solutions (refer to the Discussion).

If this test indicates your unknown is a monosaccharide, proceed to Seliwanoff's test; if it indicates your unknown is a disaccharide, proceed to the mucic acid test.

4. Seliwanoff's Test

Add 1 mL each of 0.5% solutions (dilute reagents 1 : 1) of fructose, glucose, your unknown, and 1 mL of distilled water to four separate labeled test tubes. Then add 9 mL of Seliwanoff's reagent to each test tube and heat them in a boiling water bath. Compare the times required for color development in the knowns and the unknown. Ketoses react readily, whereas aldoses react slowly. The distilled water serves as a blank. A deep red color is positive for fructose.

5. Mucic Acid Test

Place 450 mg of glucose, of lactose, and of your unknown in three separate labeled 100-mL beakers. Add 35 mL distilled water, followed by 15 mL of concentrated HNO_3 (**Caution!**) to each. Mix well and heat on a controlled hotplate or steam bath in the hood until the solutions are concentrated to approximately one-third their original volumes. Then add 10 mL distilled water to each, mix well, cover with watch glasses, and set aside in a cool place until the next laboratory period. Examine for presence of colorless crystals of mucic acid, which would be a positive test for lactose (see Discussion).

The Report form must contain all the tests, results, and conclusions you have made together with the number of your unknown and your conclusion as to its identity.

NAME _____ SECTION _____ DATE _____

Unknown Number _____ Identity of Unknown _____

IDENTIFICATION OF UNKNOWN

Test	Result	Conclusion

NAME _____ SECTION _____ DATE _____

Answer the following questions before beginning the exercise:

1. Carbohydrates comprise *monosaccharides*, *disaccharides*, and *polysaccha-rides*. Give an example of each.

2. Define the terms *monomer* and *polymer:*

3. What is the so-called building block of starch?

4. Sugars are known chemically as *polyhydroxy aldehydes* or *ketones*. Write a structure for each of these; identify the hydroxy and aldehyde/ketone groups.

5. What is meant by a *reducing* sugar?

6. Sucrose (table sugar) is a disaccharide composed of glucose and fructose com-bined chemically. It is *not* a reducing sugar. Based on this fact, what can be said about the linkage between glucose and fructose (see Fig. 39.1) in sucrose?

339

Determination of the Average Molecular Mass of a Fat

OBJECTIVE

To become familiar with a member of a class of organic molecules known collectively as lipids.

DISCUSSION

Lipids constitute another major group of biologically important organic molecules (see Exercise 39, Carbohydrates). Like carbohydrates, lipids are composed principally of carbon, hydrogen, and oxygen, but may also contain other elements; namely, phosphorous and nitrogen. Unlike carbohydrates, however, lipids contain a much smaller proportion of oxygen. The lipids are distinguished as a *single class* by their exceptionally high solubility in nonpolar solvents and, conversely, their low solubility in water. They form very high molecular mass structures, not by covalent linkages as in the case of carbohydrates, proteins, and nucleic acids, but by noncovalent, mainly hydrophobic interaction.

The major classes of lipid monomers are (a) esters of glycerol with long straight-chain fatty (aliphatic, or carboxylic) acids or the corresponding aldehydes; (b) esters of alcohols (for example, glycerol) with the long-chain fatty acids; (c) esters of sphingosine (an amino alcohol) with the long-chain fatty acids; and (d) sterols, represented by cholesterol (see Fig. 40.1). A great number of variations can be permuted. If, for example, one of the — OH groups of glycerol is phosphorylated, the group of lipids that can be formed is known as phosphoglycerides; the fatty-acid components of these may also vary. The possibilities of variation are again multiplied when the bridging compound is not glycerol, or a glycerol-related compound, but sphingosine, and so on. The functions of lipids depend on their structure. Some functions of lipids are food storage and membrane-structure formation.

Each lipid molecule usually has at least two long-chain fatty acids (most often containing 14–24 C atoms) linked together at one end by covalent bonds to a bridging compound. The fatty acids may be saturated or unsaturated (that is, contain double bonds). The unsaturated fatty acids melt, on the average, at a temperature 50°C or more below that of the corresponding saturated acids; many are liquid and are responsible for the fluidity of plant and animal oils. In the case of *fats* (neutral fats, or triglycerides), the bridging compound is glycerol, a trihydroxy alcohol, and all three of its — OH groups are esterified by fatty acids (see Fig. 40.2).

Since fats are triglycerides they can be hydrolyzed (degraded) into glycerol and fatty acids. From the quantity of KOH used in the hydrolysis, one can calculate the average molecular weight of the fat; 3 moles of KOH are needed to hydrolyze (saponify) 1 mole of a triglyceride.

$$CH_3(CH_2)_{16}COOH \qquad\qquad CH_3(CH_2)_7CH=CH(CH_2)_7COOH$$

Stearic acid
(A C-18 saturated fatty acid)

Oleic acid
(A C-18 unsaturated fatty acid)

(a) Long straight-chain fatty acids

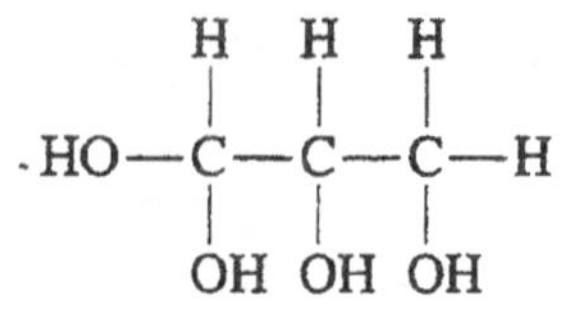

(b) Glycerol

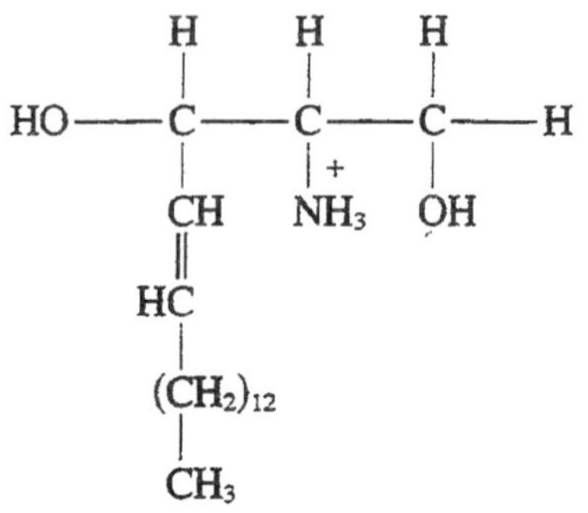

(c) Sphingosine

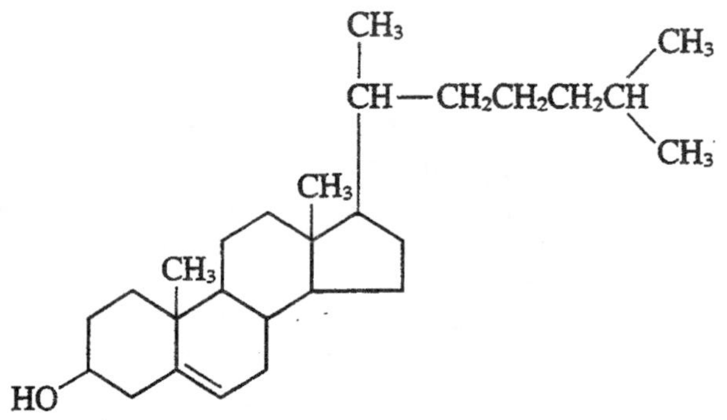

(d) Cholesterol

Figure 40.1 Some components of lipid monomers.

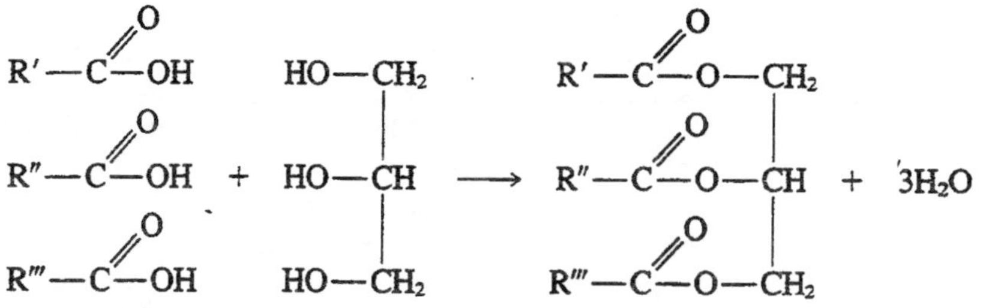

Fatty acids   Glycerol   Fat
(Triglyceride, an ester)

Figure 40.2 Formation of a fat (triglyceride) monomer from three fatty acids and glycerol. The primes (') on the R groups of the fatty acids indicate they may not be identical molecules.

PROCEDURE

In this exercise you will determine an average molecular mass for a fat by hydrolyzing a known amount of the fat with a premeasured amount of standard KOH in ethanol (the ethanol is required because a fat is insoluble in water). The amount of KOH required for the hydrolysis is found by titration with standard HCl of the excess remaining after the fat has been hydrolyzed.

If necessary, review the principles of acid–base titration (see Exercise 24) before beginning this exercise.

Weigh about 0.50 g of fat to 0.01 g into a 200-mL round-bottomed flask by difference (*1*). To the flask add 25.0 mL (measured from a buret) of standard 0.50 N KOH in 95% ethanol, and 25 mL of 95% ethanol. Attach a water-cooled condenser to the flask (see Fig. 40.3). If heating mantles are available, substitute one for the Bunsen burner. Have the instructor check your setup. Heat under reflux gently for 30 minutes. Allow the mixture to cool.

When the reflux assembly is cool, run a *small* amount of water down the condenser, and then remove the condenser from the flask. Titrate the contents of the flask with standard 0.50 N HCl, using phenolphthalein as the indicator. Record the volume of acid used in the titration (*2*).

Calculate the average molecular mass of the fat (see last paragraph of the Discussion) (*3*). Com-

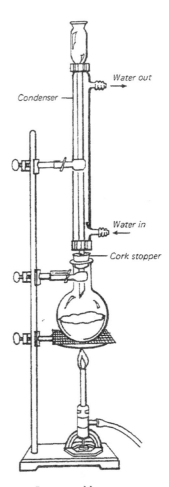

Figure 40.3   A reflux assembly.

ment on the success of your determination (*4*). Write an equation for the saponification (hydrolysis in an alkaline solution) of a typical fat (*5*).

NAME _____ SECTION _____ DATE _____

(1) Weight of flask + fat .... _____ g

    Weight of flask .... _____ g

    Weight of fat (by difference, to 0.01g) ......:................ _____ g

(2) Final buret reading .... _____ mL

    Initial buret reading .... _____ mL

    Volume of 0.50 N HCl .... _____ mL

(3) Average molecular mass of fat (based on 3 mol KOH =

    1 mol of a triglyceride (show calculations) .... _____ amu

(4)

(5) Equation:

345

NAME _____ SECTION _____ DATE _____

Answer the following questions before beginning the exercise:

1. List the major classes of *lipids*.

2. What is the one property that all lipids have in common?

3. What is usually meant by the term *fat?*

4. What is the basic reaction on which the method of determination of the average molecular mass of a fat depends?

41

Paper and Thin-Layer Chromatography

OBJECTIVE

To become familiar with the principles and practice of paper and thin-layer chromatography.

DISCUSSION

Liquid chromatography is a highly efficient means of separating the components of mixtures, even when the boiling points are so close that fractional distillation becomes impractical. Essentially, chromatography amounts to a very large number of extractions (removals) of solutes between a stationary liquid and a moving liquid. In *paper* chromatography, the stationary liquid is water, held to the cellulose of the paper by hydrogen bonding. A variety of liquids can be used for the moving phase. As the moving phase passes the point at which the solute sample has been deposited, it dissolves the components more or less readily, depending on their solubilities, and carries them along. These components are extracted back and forth between the two liquid phases; those that are more strongly held by the water film do not move as far as those more soluble in the moving film. Sharp separations can usually be achieved by proper choice of the two phases.

Thin-layer chromatography, in which a plastic or glass surface is coated with cellulose, silica gel, polyamides, or other materials, is a recent modification. It achieves sharper separations in a shorter time and permits detection of very small quantities.

A mixture to be resolved by paper or TLC (thin-layer chromatography) is placed as a small spot on one end of a strip of paper or TLC film, and solvent is allowed to move by capillary action through the spot and up the paper. The ratio of the distance traveled by a compound to that traveled by the solvent is called the R_f value:

$$R_f = \frac{\text{distance the compound traveled}}{\text{distance the solvent traveled}}$$

The R_f value of a compound is a characteristic of the compound *and the solvent used*, and serves to identify each component of the mixture.

PROCEDURE

You will be assigned either Part 1A or 1B of this exercise by your instructor, and Part 2.

1. Paper Chromatography

A. Plant Pigments. Grind some green or spinach leaves with sand in a clean mortar. The sand helps break the cell walls and release their contents. Add 4 mL of the supplied methanol in ligroin solution, and continue to grind for a short time. (**Caution: Ligroin is highly flammable; extinguish all flames.**) Decant the extract into a test tube and cork immediately. Clean the mortar.

Cut an 18-cm length of chromatographic paper from a 1.3-cm-wide roll (Whatman No. 1). Fold lengthwise without contact with the fingers, to stiffen the paper; this can be done by folding the strip inside a piece of notebook paper. Using a melting-point capillary, make a line with the plant extract about 1 cm from one end of the strip. Allow to dry. Repeat the application and drying for a total of about 10 times to ensure a heavy deposit of plant pigment.

Clamp a dry 25 × 200-mm test tube at a very slight angle from the vertical and add enough ligroin to fill the curved part of the bottom of the test tube. *The level must not be above the pigment line.* Use a long pipet and rubber bulb to avoid getting the ligroin on the test tube walls. Place the prepared strip (but do not touch the sample), sample side down (Fig. 41.1), in the test tube so that one end is immersed in the ligroin and the other end rests against the side of the test tube. Quickly stopper the tube and describe the development of the chromatogram (*1*). When the highest color zone has traveled the length of the strip, pull the strip out and dry it. Account for its appearance (*2*). Why do these pigments migrate at different rates (*3*)? Dry the strip and attach it to your report (*4*).

B. Food Colors. Cut two pieces from a 1.3-cm-wide role of Whatman No. 1 chromatographic paper, long enough to extend to the bottom of a 500-mL wide-mouth Erlenmeyer flask and lap over the rim enough to be fastened in place with Scotch tape (see Fig. 41.2). Straighten the chromatographic tapes from curling (use clean, dry hands). Draw a light *pencil* (not ink) line about 1.5 cm from the bottom of the paper, parallel to the short side.

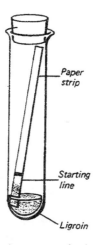

Figure 41.1 Paper chromatography for plant pigments.

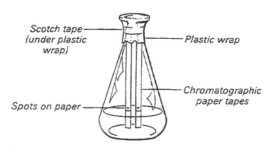

Figure 41.2 Paper chromatography for food colors.

A commercial food color set contains red, blue, green, and yellow samples. Using separate micropipets, or melting-point capillaries, place two not-too-large spots just above the pencil line on each tape, approximately 0.5 cm apart—a total of the four different colors for the two tapes. Label each spot with a light pencil mark next to the spot just above the pencil line. Allow the spots to dry.

Fasten the tapes in the Erlenmeyer flask, with the end of the spotted tapes just touching the bottom of the flask. Make the tapes hang as vertically as possible.

Dilute 10 mL of 2-propanol with 5 mL of water in a small beaker and transfer this solution to the bottom of the flask by pouring it through a long-stemmed funnel, without splashing. The level of the solvent must *not* be above the spot. Cover the mouth of the flask with plastic wrap. Allow the chromatogram to develop; discontinue before the solvent reaches the top of the filter paper (about 30 minutes).

Remove the tapes, mark the top of the solvent front while it is still visible with a pencil line, and allow it to dry. Circle the individual spots.

Suppose one of the colors, after chromatogramming, gave rise to two separated substances (see Fig. 41.3). The R_f values would be calculated as follows:

$$R_{f_1} = \frac{\text{distance from initial spot to midpoint of spot 1}}{\text{distance of movement of solvent front}}$$

$$R_{f_2} = \frac{\text{distance from initial spot to midpoint of spot 2}}{\text{distance of movement of solvent front}}$$

In this manner, calculate the R_f values (see Discussion) of *each* separated color of each sample of food color (*5*).

Attach the tapes to your report (*6*).

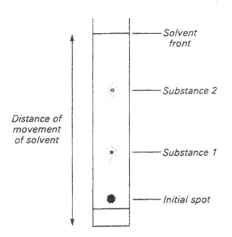

Figure 41.3   A hypothetical developed chromatogram.

2. Thin-Layer Chromatography*

Obtain a 5 × 10-cm section of coated TLC film. Draw a *light pencil* line 1 cm from the 5-cm end. Make four spots just above the pencil line, one for each of the four different phenol samples† supplied for this experiment, using a separate micropipet or melting-point capillary for each. Make a fifth spot from the supplied mixture containing all four phenols. Be careful not to disturb the coating on the film when applying the samples. The spots will be colorless. Label each spot with a light pencil mark next to the spot.

Chromatogram the film in a 400-mL beaker containing 6 mL of the supplied developing solvent until the solvent has advanced to within 2 cm of the top of the film (about 30 minutes). Remove the film, mark the upper limit of the solvent front with pencil, and dry.

Transfer the dry film into a clean, dry 400-mL beaker with a *small* crystal of iodine. Cover the beaker with a watch glass, and warm gently for a minute or so, preferably on a hot plate at the warm setting (**use the hood**). Iodine vapor absorbed by the phenols will reveal the location of each.

Remove the chromatogram from the beaker. Carefully outline each darkened spot, mark its midpoint, and measure the distance of the midpoint of each spot to the starting line. Determine the R_f values, and record in the Report form (7). Compare your values with the average values given (8).

On the fifth developed composite spot, relate the separated components with the four individual spots (9). What can be said about the separation of compounds with similar R_f values (10)? Suggest a way this problem could be solved (11).

Attach your chromatogram to the Report form (12).

---

*Part 2 of this exercise was adapted from *Organic Experiments*. 6th ed., by Linstromberg and Baumgarten (Heath, Lexington. Mass.: 1983).
† Solutions are 6% w/v in ethanol.

#1 (Phenol)

#2 (Catechol)

#3 (Resorcinol)

#4 (Pyrogallol)

NAME _____   _____ SECTION _____ DATE _____

1. Paper Chromatography

A. PLANT PIGMENTS

(1)

(2)

(3)

(4) Attached chromatogram (use Scotch tape on top)

B. FOOD COLORS

(5)

Food color	No. of components	Color of components	R_f values
Red			
Blue			
Green			
Yellow			

(6) Attached chromatogram tapes (use Scotch tape on top)

2. Thin-Layer Chromatography

(7)

Compound	R_f values (average)	R_f values (experimental)
Phenol	0.84	
Catechol	0.72	
Resorcinol	0.77	
Pyrogallol	0.53	

(8)

(9)

(10)

(11)

(12) Attached chromatogram (use Scotch tape on top)

NAME _____ SECTION _____ DATE _____

Answer the following questions before beginning the exercise:

1. What is meant by "TLC"?

2. List some of the advantages of TLC over paper chromatography.

3. Define the term R_f value.

4. Explain how R_f values can be used in *qualitative* analysis.

.5. If you change the *solvent* system of a chromatographic analysis, what happens to the particular R_f values for the components separated?

6. Name some characteristics of compounds that determine how far the components travel in a solvent system.

7. Suggest a way that TLC can be used as a *quantitative* analysis.

8. Iodine *sublimes*. What does this mean?

42

Amino Acid Identification by Chromatography

OBJECTIVE

To become familiar with the principles and practice of techniques of chromatography in the identification of amino acids.

DISCUSSION

Separating, isolating, and identifying amino acids obtained by hydrolysis of proteins has been simplified enormously through the use of chromatography. A solution of ninhydrin in ethanol is used to develop color to identify the separated amino acids, and makes it possible to follow the progress of protein hydrolysis. Ninhydrin releases the alpha—NH_2 group from amino acids and forms colored compounds with the NH_3 produced. In a protein, practically all the amino nitrogen is chemically bound up in the peptide bonds

$$
\begin{array}{cc}
O & H \\
\parallel & \mid \\
-C- & N- \\
\end{array}
$$

and gives no color with ninhydrin. However, on hydrolysis, alpha—NH_2 groups are made available for color production with ninhydrin.

The following flow equations illustrating the oxidative deamination of amino acids show the stages of the ninhydrin reaction, producing NH_3 (ammonia), which forms the violet pigment with ninhydrin:

$$
\underset{\text{R—CHCOOH}}{\overset{NH_2}{\mid}} \xrightarrow{[O]} \underset{\text{R—CCOOH}}{\overset{NH}{\parallel}} \xrightarrow{H_2O}
$$

$$
NH_3 + \underset{\text{RCCOOH}}{\overset{O}{\parallel}} \longrightarrow RCHO + CO_2
$$

The overall reaction is

Ninhydrin   Amino acid

Violet pigment

+ $RCHO$ + CO_2 + $3H_2O$

Amino acids can be identified on the basis of their R_f values; this is the distance a particular amino acid moves divided by the distance the solvent front moves. Since amino acids may have one amino group to one carboxylic acid group, or two to one, or one to two, their migration rates depend on the pH of the developing solvent as well as molecular mass.

PROCEDURE

You will be assigned Part 1, 2, or 3 by your instructor.

1. Ascending-Paper Chromatography

Place 10 mL of 2% aqueous ammonia and 20 mL of 2-propanol in a 1-L beaker. Obtain five melting-point capillaries, and sufficient aluminum or plastic foil to cover the beaker.

Label five 10-cm test tubes *Leucine, Tyrosine, Aspartic acid, Glycine,* and *Unknown*. Put a few drops of each amino acid (0.05 M in 0.05 M HCl) and the unknown in the corresponding test tube. Place a capillary tube in each sample. Obtain a 12 × 22-cm sheet of chromatographic paper. Being careful not to transfer grease from your fingers on the sheet, draw a *pencil* line 2 cm from the bottom on the long dimension.

Along this line make ten dots 2 cm apart; below the first five dots write the first letter of the amino acid (including U for unknown) that is to be placed there. Below the last five dots write the first letter of the acid but in an order different from the first. Each amino acid and the unknown will have two dots.

With the capillaries, put a small amount of each acid and the unknown on the first five correspond-ing dots; the spots should not be more than 3 mm in diameter. Let dry in air. Repeat the application once more to the remaining five dots and dry again. Make a cylinder out of the paper, sample side out, and staple so that the ends do not touch.

Insert the cylinder into the 1-L beaker without touching its sides and cover the beaker tightly with the aluminum or plastic foil. Do not splash. Allow the solvent to rise nearly to the top of the cylinder; this may take 90 minutes. Remove the cylinder, mark the solvent front, and place upside down on a sheet of paper until most of the solvent has evaporated. Open the cylinder and support it (**in the hood**) in such a way that it can be sprayed. Gummed tape or paper clips and a short length of string can be used for this purpose.

Spray the paper chromatogram with 2% ninhydrin in ethanol, **being careful not to inhale the spray or let it touch the skin.** Allow to dry. If necessary, the experiment can be interrupted at this point. Complete the drying in a 105°C oven for 15 minutes.

Draw an X through the center of each spot. Calculate the R_f value for each spot (*1*). Identify the components of the unknown by comparing its R_f values with R_f values of the known amino acids (*2*) (see instructor). Would the R_f values you determined be the same for a different solvent system

(3)? Would the speeds of migration of these amino acids be in the same order with a different solvent system (4)? Account for the relative positions of the dots you obtained on the basis of the structural formulas of the four amino acids (5). Attach the filter-paper sheet to your report (6).

2. Radial-Paper Chromatography

It is instructive to compare the ascending-paper chromatographic technique with a simpler one, in which the sample migrates horizontally outward from the center of a circular filter paper.

Obtain a Petri dish and a circle of suitable chromatographic paper slightly larger than the dish. Find the center of the paper and draw through it, with pencil, five lines dividing the paper into approximately equal zones. Draw a circle 1.5 cm in diameter in the center of the paper. In each sector, put the first letter of the amino acid to be applied in that sector; if the same amino acids are to be used as in the preceding experiment, the letters will be *L*, *A*, *T*, *G*, and *U* for unknown. Cut a small hole in the center of the paper for insertion of a wick.

Set up the test tubes, capillaries, and samples as in Part 1. Apply the samples as before, but put each on the penciled circle in its proper sector; two applications of each sample should be made. The wet spots should not spread more than 3 mm. Finally, dry in a 105°C oven for 10 minutes. Meanwhile, half fill the lower half of the Petri dish with a mixture of 1 volume of 2% aqueous ammonia and 2 volumes of 2-propanol; close the dish with the upper half.

When the paper is dry, open the dish and place the paper on the bottom half of the dish. Make a wick from a small piece of rolled up filter paper and insert it through the center hole into the solvent. Replace the upper part of the dish and allow the solvent to advance in the paper nearly to the rim of the dish.

Remove the paper, mark the position of the solvent front, and dry at 105°C for 5 to 10 minutes. Hang the paper in the hood and spray lightly with 2% ninhydrin in ethanol, **being careful not to inhale the spray or allow it to touch the skin.** Dry again at 105°C. Mark the center of the arc obtained for each amino acid. Calculate the R_f value for each amino acid (7). Compare these R_f values with the values obtained by ascending chromatography in Part 1 (8). Comment on any differences in the shapes and sizes of the spots obtained by the two methods (9). What are the advantages and disadvantages of each method (10)? Attach the filter paper to your report (11).

3. Thin-Layer Chromatography

The developing chamber is a jar about 12 cm high, equipped with a metal screw-top lid. The lid has two parallel slots, 6 cm long and 2 cm apart. Pour the top layer of the developing solution (a 1-butanol–acetic acid–water mixture) into the jar to a depth of 1 cm; slosh it around the sides. Replace the lid and allow to stand 30 minutes for equilibration.

Meanwhile obtain two commercial cellulose-coated plastic chromatographic strips measuring 5×13 cm. Handle only by the edges. Ask the instructor if they need activation. Label four small test tubes *Alanine*, *Leucine*, *Serine*, and *Valine*. Obtain 4 capillaries as in Part 1 and insert one of them in each test tube. Put a few drops of the corresponding 0.05 M solutions into the test tubes.

Using the capillaries, apply samples of the amino acids to the strips—two samples per strip. The spots should be 2 cm from the bottom of the strips and 2 cm apart; the wet spots should not be more than 3 mm in diameter. Suspend the strips and dry thoroughly with a warm-air blower or infrared lamp. Record the location of each amino acid sample (12).

Lower the strips through the slots in the developing jar lid so the lower ends are immersed; do not open the lid. Secure the upper ends with paper clips. Allow the solvent to rise 10 cm; this may take up to 2 hours. Remove the strips, mark the solvent front, and hang **in the hood** to dry. Spray with ninhydrin solution, made by mixing 50 mL of Visualizer Solution I with 3 mL of Visualizer Solution II just before use. **Be careful not to inhale the spray or allow it to touch the skin.** Place the strips on a glass plate and heat in a 105°C oven for 15 minutes. Mark the locations of the spots. Calculate the R_f value for each amino acid (13). Comment on the success of the separation (14). Attach the strips to your report (15).

If the spots are too faint to locate, the experiment should be repeated with another strip, making several applications of sample. Dry between applications. If the spots are so large that they overlap badly, too much sample has been applied.

Optional: The instructor may assign you an unknown. This will be one or more of the four amino acids above. Since the conditions may not be exactly the same as before, you should run one of these amino acids along with the unknown on the same strip. Report the components of the unknown (*16*). Attach the strip to your Report form, after marking the solvent front and the spots (*17*).

NAME _____ SECTION _____ DATE _____

Amino Acid Identification by Chromatography

(1), (2), (7), (8)

Amino acid	R_f (calculated) ascending method	R_f (calculated) radial method	R_f (known)
Leucine			
Tyrosine			
Aspartic acid			
Glycine			
Unknown			

(3)

(4)

(5)

(6) Attached chromatogram (use Scotch tape on top):

(9)

(10)

(11) Attached chromatogram (use Scotch tape on top):

(12), (13)

Amino Acid	R_f (calculated)
Alanine	
Leucine	
Serine	
Valine	

(14)

(15) Attached chromatogram (use Scotch tape on top):

Optional:

(16)

(17) Attached chromatogram of unknown (use Scotch tape on top):

NAME _____ SECTION _____ DATE _____

Answer the following questions before beginning the exercise:

1. What is the type formula for an amino acid?

2. What two functional groups are present in an amino acid?

3. What is the *peptide bond?*

4. What are the products of hydrolysis of proteins?

5. Why is ninhydrin spraying needed in the chromatogramming of amino acids?

6. Name two factors that influence the migration rates of amino acids (see Discussion).

7. Explain how R_f values for amino acids are obtained.

43

Spectrophotometric Determination of the Equilibrium Constant of a Coordination Complex

OBJECTIVE

To become familiar with the application of absorption spectrometry to the determination of an equilibrium constant. (Review Exercise 20.)

DISCUSSION

When white light is passed through a colored solution, certain wavelengths are absorbed. The absorbed wavelengths are complementary to the color observed; for example, $FeSCN^{2+}$ appears orange because it absorbs blue-green. The amount of absorption is proportional to the concentration of the absorbing species and is measured most accurately by electronic means. The spectrophotometer can be set to pass a narrow band of wavelengths, those corresponding to strong absorption of the substance analyzed; for $FeSCN^{2+}$ the best wavelength is 450 millimicrons (nanometers or nm), that of blue-green light (1 millimicron = 10^{-7} cm = 10^{-9} m = 1 nm). The light passing through the sample strikes a photocell; the effect is amplified and read on a scale.

PROCEDURE

In this exercise you will study the following equilibrium:

$$FeSCN^{2+} \rightleftharpoons Fe^{3+} + SCN^-$$

This equilibrium may be written more exactly as

$$Fe(H_2O)_5SCN^{2+}(aq) \rightleftharpoons$$
$$Fe(H_2O)_6^{3+}(aq) + SCN^-(aq)$$

The spectrophotometer reads directly in a logarithmic absorbance scale, and for a limited range of concentration, $A = k[FeSCN^{2+}]$. (Although the absorbing species is more complicated than indicated by the formula $FeSCN^{2+}$, this is not a matter of concern for the purposes of this exercise. The hydrated Fe^{3+} ion does not absorb at the same wavelength.) You will first determine the absorbance for a known concentration of $FeSCN^{2+}$ to evaluate k. Subsequently you will calculate the concentrations of $FeSCN^{2+}$ at various dilutions of the sample from k and A. The equilibrium constant is then determined from this information.

Rinse a clean Erlenmeyer flask with a 10-mL portion of standard 5×10^{-2} M $Fe(NO_3)_3$. After draining, add about 50 mL of this solution to the flask. Record the concentration to three significant figures (*1*). Similarly rinse another Erlenmeyer flask with standard 5×10^{-4} M KSCN and add 50 mL. Record its concentration to three significant figures (*2*). (Both solutions are 0.5 M in nitric acid to avoid hydrolysis.)

Before proceeding, it is advisable to see Part 3G for information regarding the use of volumetric glassware. Attach a safety pipet filler to a 10-mL volumetric pipet and rinse twice with small portions of the $Fe(NO_3)_3$ solution. Deliver 10 mL of the $Fe(NO_3)_3$ solution into a 25-mL volumetric flask. Add about 5 mL of 2.5 M HNO_3 and dilute *exactly* to the etched line. Label the flask *Dilution 1*. Hold the ground-glass stopper securely in the volumetric flask, and thoroughly mix the contents by inverting the flask several times.

Prepare Dilution 2 by pipetting 10 mL of Dilution 1 into another 25-mL volumetric flask, adding 5 mL of the nitric acid, and diluting to the mark. Label the flask *Dilution 2*. Mix the contents.

Prepare Dilution 3 by pipetting 10 mL of Dilution 2 into another 25-mL volumetric flask, adding 5 mL of the nitric acid, and diluting to the mark. Label the flask *Dilution 3*. *In each case the pipet must be rinsed with the solution to be used next.* Calculate the concentrations of $Fe(NO_3)_3$ in Dilutions 1, 2, and 3 (*3*).

Rinse the pipet with water and then with small portions of the KSCN solution. Measure four 10-mL samples of the KSCN solution into dry Erlenmeyer flasks labeled *A*, *B*, *C*, and *D*. Rinse the pipet with water and then with the *undiluted* $Fe(NO_3)_3$ solution. Deliver 10 mL of the undiluted solution into flask A. Following the same procedure, add 10 mL of Dilution 1 into flask B, and so

on. *Always rinse the pipet with the solution to be used next.*

Ignoring reaction, calculate the concentrations of $Fe(NO_3)_3$ and KSCN in flasks A, B, C, and D (*4*). Keep in mind that doubling the volume drops the concentration to half. These samples are to be analyzed spectrophotometrically at a wavelength of 450 nm. The Spectronic 20 is a suitable instrument, although several others are also satisfactory. See the Procedure in Exercise 32 for directions for the Spectronic 20. If another instrument is to be used, see your instructor.

Always rinse the sample cuvette with the next sample to be determined.

Determine the absorbances of samples A, B, C, and D (*5*).

Calculate the proportionality constant k, in the formula $A = k[FeSCN^{2+}]$, from the absorbance of Sample A; assume that the concentration of Fe^{3+} is in such excess that almost all of the SCN^- is converted to $FeSCN^{2+}$ (*6*).

Using the calculated value of k and the absorbances, compute the concentration of $FeSCN^{2+}$ in Samples B, C, and D, and the equilibrium constant K for each sample (*7*). Calculate the average value of the equilibrium constant K (*8*).

For Sample A it was assumed that the concentration of SCN^- is negligible. How well is the assumption justified (*9*)?

Suggest some principal sources of error in this method of determining the equilibrium constant (*10*).

NAME _____ SECTION _____ DATE _____

(1) $[Fe(NO_3)_3] =$ _____ M

(2) $[KSCN] =$ _____ M

(3) Calculate:

Dilution	$[Fe(NO_3)_3]$
1	
2	
3	

(4), (5) Data:

Sample	$[Fe(NO_3)_3]$	$[KSCN]$	Absorbance (450 nm)
A			
B			
C			
D			

(6) $k =$ _____ (Show calculation below)

(7) Calculate:

Sample	$[FeSCN^{2+}]$	Equilibrium constant K
B		
C		
D		

(Sample calculation)

(8) Average equilibrium constant $K = $ _____

(9)

(10) Sources of error:

NAME ——————————————————— SECTION ————— DATE —————

Answer the following questions before beginning the exercise:

1. Write the equation for the K_{eq} of the reaction being studied in this exercise.

2. What is the substance that is being measured spectrophotometrically?

3. What color does this substance (see Question 2) have (see Discussion)? Explain why.

4. In the equation, $A = k[FeSCN^{2+}]$, what is k?

369

44

Conversion of Aluminum Scrap to Alum

OBJECTIVE

To convert a sample of "scrap" aluminum to alum; to demonstrate that a valuable resource such as aluminum can be conserved by recycling to produce a useful substance.

DISCUSSION

Aluminum is the most abundant metal and the third most abundant element in the earth's crust. Our society has used aluminum to manufacture everything from lawn chairs to beverage cans to building materials. The metal is particularly useful because of its resistance to corrosion.

Approximately half the aluminum produced domestically is converted to alloys that are nearly as strong as steel but are much lighter in weight. One popular alloy, Duralumin, contains small amounts of copper, magnesium, and manganese, and is used extensively for manufacturing lightweight structural components, such as those needed in the aircraft industry. Most alloys are coated with a thin layer of aluminum to provide protection for the less resistant components. The surface of the thin aluminum layer is spontaneously oxidized to form an oxide coating that is virtually impervious to attack by atmospheric oxygen and water.

Metallic aluminum, Al, and aluminum oxide, $Al_2O_3 \cdot 2H_2O$, readily react with concentrated alkaline solutions to form aluminates according to the following reaction:

$$2Al(s) + 2KOH\ (aq) + 6H_2O(l) \longrightarrow 2KAl(OH)_4(aq) + 3H_2(g) \qquad (1)$$

$$Al_2O_3 \cdot 2H_2O(s) + 2KOH(aq) + H_2O(l) \longrightarrow 2KAl(OH)_4(aq) \qquad (2)$$

The aluminates are soluble but alloy elements or impurities remain undissolved.

The aluminates can be refined to produce metallic aluminum through an electrolytic process or other useful aluminum compounds. As an example, the sulfates are the cheapest soluble salts of aluminum and are widely used in the purifying of water, manufacturing of fabric dyes, sizing of paper, and waterproofing of fabrics.

"Alums" are double salts, referred to simply as *alum;* for example, $KAl(SO_4)_2 \cdot 12H_2O$. Two different cations are present in the formula.

PROCEDURE

In this exercise you will study a method for converting "scrap" aluminum, such as that from beverage cans or food trays, to an alum. Upon cooling a solution containing the ions of potassium and aluminum and sulfate ions, $KAl(SO_4)_2 \cdot 12H_2O$ will crystallize out in good measure. The alum, however, is slightly soluble in water at low temperatures, and a small amount will remain in solution..

The preparation of alum from aluminum metal follows the stepwise sequence shown:

Aluminum (Metal) $\xrightarrow{1}$ Aluminates $\xrightarrow{2}$
Aluminum ions $\xrightarrow{3}$ Alum

Aluminum metal must be reacted (digested) with potassium hydroxide to produce an aqueous solution of $KAl(OH)_4(aq)$. **Caution: This reaction generates hydrogen gas, and should be done in a well-ventilated open area or under a hood. No flames should be nearby** (see Eq. 1). Obtain approximately 2 g of aluminum scrap, which has been cut into small pieces, and place the scrap into a 250-mL beaker. Record the mass of the aluminum scrap (nearest 0.01g) (1) used on the Report form. Add 60 mL of 1.5 M KOH to react with the metal. *Gently* heat the mixture on a hot plate (no flame) until the pieces of scrap are no longer visible, about 20 minutes. Considerable bubbling or frothing may occur during the reaction. Do not allow the mixture to spill over the side of the beaker.

After the metal has been digested, slowly filter the hot solution, using a funnel containing a *small* plug of glass wool, into a 150-mL beaker. Allow the solution to cool. The aluminate ions, $Al(OH)_4^-$, are converted to aluminum ions by reaction with hydrogen ions from sulfuric acid:

$$Al(OH)_4^-(aq) + H^+(aq) \longrightarrow Al(OH)_3(s) + H_2O(l) \quad (3)$$

Aluminum hydroxide, $Al(OH)_3(s)$, will develop during the addition of acid and appear as gelatinous, white lumps. Further addition of acid will cause the hydroxide to react and dissolve according to the following reaction:

$$Al(OH)_3(s) + 3H^+(aq) \longrightarrow Al^{3+}(aq) + 3H_2O(l) \quad (4)$$

Converting aluminate ions to aluminum ions should require about 20 mL of 9 M H_2SO_4 solution. Slowly add the H_2SO_4 solution, with stirring, to the aluminate solution; the $Al(OH)_3(s)$ should disappear. If some precipitate remains, gently heat the mixture, with stirring, until all the $Al(OH)_3$ dissolves. Sulfuric acid has a dual function in the reaction process; it provides sulfate ions for the production of alum along with the required hydrogen ions.

Crystals of alum will form as the clear solution cools. Allow the clear, hot solution to cool for a few minutes, and then place it in an ice bath. After about 20 minutes in the ice bath, crystals of alum should be visible in the bottom of the beaker. If not, stir the solution gently and scratch the inside wall of the beaker with a glass rod until crystals form. Alum, a double salt, is produced from the ions according to the following equation:

$$K^+(aq) + Al^{3+}(aq) + 2SO_4^{2-}(aq) + 12H_2O(l) \longrightarrow KAl(SO_4)_2 \cdot 12H_2O(s) \quad (5)$$

Set up a suction filter apparatus as shown in Fig. 44.1. (Further information about the suction filtration method is found in Part 3, page 390.) Preweigh the filter paper, record the mass (2), assemble it in the Büchner funnel, turn on the suction, and proceed to collect the alum crystals on the paper. The mixture should be swirled to get the crystals in suspension. Then slowly pour the mixture into the funnel.

After the liquid has drained completely from the crystals, extract any residual water by washing the

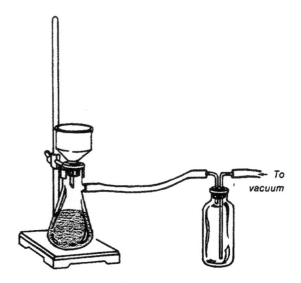

Figure 44.1 Suction filtration apparatus.

crystals twice with ethanol: Add 15 mL of ethanol, filter dry, and repeat. Allow the suction to continue for 5 minutes to air-dry the crystals. To complete the drying process, place the crystals and filter paper into a clean, dry beaker and store them in a dry place until your next laboratory period.

Weigh the filter paper–alum combination, record the mass (3), subtract the mass of the filter paper, and record the mass of alum produced (4). Calculate the percentage conversion of aluminum to alum based on Eq. (5). Aluminum ions in the equation can be considered as aluminum metal. Assume the scrap to be pure aluminum for the calculations. Record your calculations on the Report form (6).

NAME _____ SECTION _____ DATE _____

(1) Mass of aluminum scrap used .... _____ g

(3) Mass of filter paper + KAl(SO$_4$)$_2$ · 12H$_2$O .... _____ g

· (2) Mass of filter paper .... _____ g

(4) Mass of KAl(SO$_4$)$_2$ · 12H$_2$O (3) − (2) .... _____ g

(6) Percentage yield of KAl(SO$_4$)$_2$ · 12H$_2$O (Show all calculations below.)

Moles of Al scrap used .... _____ mol

Moles of KAl(SO$_4$)$_2$ · 12H$_2$O produced .... _____ mol

Moles of KAl(SO$_4$)$_2$ · 12H$_2$O predicted .... _____ mol

$$\text{Percentage yield} = \frac{\text{moles KAl(SO}_4)_2 \cdot 12\text{H}_2\text{O produced}}{\text{moles KAl(SO}_4)_2 \cdot 12\text{H}_2\text{O predicted}} \times 100$$

$$= \underline{\hspace{2cm}} \times 100 = \underline{\hspace{2cm}} \%$$

NAME _____ SECTION _____ DATE _____

Answer the following questions before beginning the exercise:

1. What is an alum?

2. An electroplating unit used for winning aluminum from Al_2O_3 draws 1.0 ampere of current at a power requirement of 0.29 kilowatt hours per day. If the unit produces 8.0 g of aluminum per day, what is the cost of producing the aluminum for a can weighing 10.0 g at 5¢ per kilowatt hour?

3. The solubility of alum, $KAl(SO_4)_2 \cdot 12H_2O$, in water solution at 0°C is approximately 114 g per liter. For this experiment calculate the mass of alum remaining in solution after completion of the crystallization based on a solution volume of 40 mL.

377

45

Analysis of Sodium Bicarbonate Samples

OBJECTIVE

To use the stoichiometry of the chemical reaction by which sodium bicarbonate (sodium hydrogen carbonate) is converted to sodium carbonate as an indirect means of calculating the amount of sodium bicarbonate in the original sample.

DISCUSSION

Sodium bicarbonate is converted to sodium carbonate by heating above 100°C, as shown in the following balanced equation:

$$2NaHCO_3(s) \xrightarrow{\Delta} Na_2CO_3(s) + H_2O(g) + CO_2(g)$$

Thus the amount of mass lost by the heated sample is equal to the mass of the water and the carbon dioxide released to the air by the thermal breakdown of the carbonic acid first formed:

$$H_2CO_3 \xrightarrow{\Delta} H_2O(g) + CO_2(g)$$

If the sample is impure, the percent of impurity can be calculated from its mass before heating. Answer the Thought Questions before calculating your experimental results.

PROCEDURE

Obtain a sample of sodium bicarbonate. Record its sample number or description of it in the Report form (1). Weigh a clean, dry test tube to the nearest 0.01 g (2). Weigh out about 5 g of the sample, using a slick weighing paper. Carefully transfer the entire sample to the weighed test tube and weigh the test tube and contents, again to the nearest 0.01 g (3). The difference between (3) and (2) is the mass of your sample (4).

Using a test tube holder, heat the sample in a Bun-

sen flame, gradually at first and then more strongly, over the length of the test tube.

Remove the test tube and contents from the flame and allow to cool, placing on a clean, dry surface such as a *clean* wire gauze. *After cooling to room temperature*, weigh and record the weight to the nearest 0.01 g (5).

Repeat the heating and cooling processes, again weighing the test tube and contents to the nearest

0.01 g (5). If the two weighings are farther apart than 0.10 g, repeat the procedure until a constant weight is obtained. The weight of your sample is (6); namely, (5) − (2). Only in this way can you be sure the reaction given in the Discussion is complete, and stoichiometry can be applied in the calculation.

Use Dimensional Analysis (see Appendix), and show your calculations (7).

Optional: Repeat the exercise with a sample of pure baking soda USP (sodium bicarbonate). Use a similar reporting form (8). Compare results (9).

NAME _____ SECTION _____ DATE _____

(1) Sample number, or description of:

Before Heating

(2) Weight of test tube .... _____ g

(3) Weight of test tube and contents .... _____ g

(4) Mass of sample (3) − (2) .... _____ g

After Heating and Cooling

	Trial 1	*Trial 2*
(5) Weight of test tube and contents	_____ g	_____ g
(6) Mass of sample (5) − (2)	_____ g	_____ g

(7) Calculations: (use dimensional analysis; see Thought Questions)

Optional:

(8)

(9) Comparison of results:

NAME _____ SECTION _____ DATE _____

Answer the following questions before beginning the exercise:

Set up these calculations in dimensional analysis form (see Appendix).

1. How many *moles* of $NaHCO_3$ are contained in 5.00 g of a pure sample?

2. Theoretically, how many *moles* of H_2O and CO_2 (H_2CO_3) would be lost by heating a 5.00-g sample of pure $NaHCO_3$? (Refer to the coefficients in the balanced equation.)

3. Convert the value obtained in Question 2 to *grams* of mass lost due to the heating.

4. Suppose a heated 5.00-g sample lost 1.00 g in mass. Calculate the percent of impurity in this hypothetical sample.

5. Can you determine the *nature* of the impurity using this method?

Part 3

Some Laboratory Skills and Devices

A. COMMON LABORATORY EQUIPMENT

To be able to identify some of the apparatus you will use in the laboratory, study the drawings on pages 12–13.

B. HEATING

The Bunsen burner, or some modification of it such as the Tirrill type (Fig. B.1), is used for most laboratory heating. The burner is a device so constructed that a gaseous fuel and air are mixed and then burned at the top of a vertical tube (the barrel). By adjusting the relative amounts of air and gas admitted, one can control the nature and temperature of the flame. The most commonly used gaseous fuel is natural gas, which is composed principally of the hydrocarbon methane, CH_4. A nonluminous flame results from the complete combustion of the fuel gas; this is the hottest flame obtainable with a burner. A luminous flame results when there is only partial combustion of the gaseous fuel, caused by an insufficient supply of air. Its appearance is due to incandescent particles of unburned carbon. The luminous flame is the coolest flame obtainable with the burner.

Examine your burner in detail before operating it; identify all its parts and their function. To operate the burner, attach it to the gas supply by a rubber tube. Turn on the gas and light the burner by bringing a lighted match or striker from the side to the top of the burner. Adjust the mixture of gas and air until the flame consists of an inner cone and a nearly colorless outer cone. If the flame should "strike back" and burn at the base of the burner, turn off the gas and let the burner cool before relighting it. Decrease the amount of air admitted to the burner or adjust the flow of gas to prevent striking back.

The relative temperatures of the different portions of the flame can be determined by holding a thin piece of copper wire across it. The hottest part of the flame is the top of the inner cone of a properly adjusted burner (Fig. B.2).

Ordinary beakers, crucibles, and other objects to be heated are placed just above the hottest portion of the flame, which is thus allowed to spread about them. If the air supply to the burner is insufficient, the resulting luminous flame will

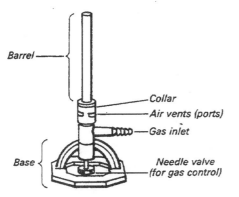

Figure B.1   The burner (Tirrill type). A primary source of heat in the beginning laboratory.

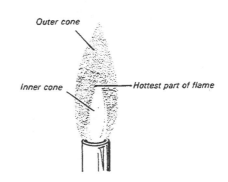

Figure B.2   The flame of a properly adjusted burner.

Figure B.3 Wing top for the laboratory gas burner.

deposit carbon on objects heated, and also will be inefficient in heating.

If a spread-out flame is needed—for glass bending, for instance—a wing top, or flame spreader, is placed on top of the burner (Fig. B.3). The wing top must not be bent, because a bent wing top results in an uneven flame that is unsuitable for glassworking. If your wing top is bent, use the end of a triangular file to make its top opening uniform along the entire length.

C. GLASSWORKING

Caution: When working with glass tubing, safety goggles must be worn.

Cutting Glass Tubing. Make a scratch across the glass tubing at the desired point with a single stroke of a triangular file. The stroke should be away from the user, with the pointed end of the file in hand. Place both thumbs close together on the side of the tubing opposite the scratch, and snap the tubing in a direction away from you against the pressure of the thumbs (Fig. C.1). If gentle pressure does not cause the glass to break at the point of the scratch, make the scratch slightly deeper and repeat the operation. Tubing must always be held in a towel while pressure is being applied to prevent injury to the hands.

Fire-Polishing Glass Tubing. All edges of glass tubing must be fire-polished to avoid cutting the hands or the rubber stopper or rubber tubing into which the glass may be inserted. Hold the sharp edges of the tubing at an angle in the flame and rotate the tubing until a bright yellow color is imparted to the flame by the glass. The result should be a smooth end. (Remember a heated piece of glass stays hot for quite some time; place it on an insulated material such as a clay triangle with the hot part away from you.)

Bending Glass Tubing. Attach a wing top to the burner to spread the flame (Fig. C.2); this makes it possible to heat a greater length of glass tubing. Hold a piece of tubing horizontally in the upper portion of this flame. Support the tube *loosely* with one hand and *rotate* it with the other to ensure uniform heating. Uniformity of heating can be judged by the yellow color imparted to the burner flame. Continue rotation until the glass softens, then remove the tubing from the flame and bend it to the desired angle. Figure C.3 shows a good bend (a) and a bad bend (b). In a bad bend (b), the glass tubing becomes undesirably constricted, a condition that can result from nonuniform heating. While the glass is cooling, it should be suspended or laid upon a piece of flame and heat-resistant material.

Preparing a Pipet with a Capillary Tip (Medicine Dropper). To make a 2-mL pipet with a capillary tip, draw out a 6-mm glass tube in the following manner: Rotate the middle 3-cm section of a 25-cm length of 6-mm glass tubing in a burner flame; do not attach the wing top. When the glass becomes soft, remove from the flame and pull evenly on the ends. Allow the tubing to cool and cut at the narrowest point. Fire-polish all sharp edges. Narrow the opening in the pointed end to 1 mm by rotating it *briefly* in the flame. Again allow cooling and then rotate the other end in the flame. When the glass is soft, remove from

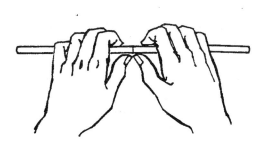

Figure C.1 Correct positioning of the fingers for breaking glass tubing (towel not shown).

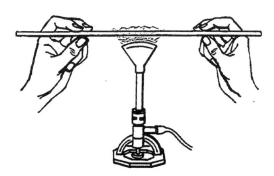

Figure C.2 Method of heating glass tubing preparatory to bending.

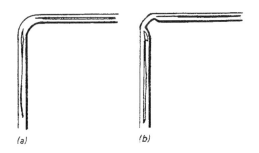

Figure C.3 Glass tubing bent at an angle of 90 degrees. (a) good bend: (b) bad bend.

the flame and insert a piece of wood or carbon shaped in a pencil sharpener in the opening to spread it slightly. Reheat the end of the tubing, continuing rotation; when the glass becomes soft, hold the tube vertically and press down on a non-flammable surface, to form a thick ring or bead at the bottom end. After the glass is cool, attach a rubber bulb and the pipet is ready to use. Squeeze bulb to expel air before inserting the pipet into a liquid.

Preparing Capillary Tubes for Melting-Point Determination. This procedure requires some practice for good results. Heat the middle 3-cm section of a 25-cm length of 6-mm glass tubing in a broad flame while rotating it evenly and continuously. When the glass is quite soft, remove it from the flame and slowly pull the ends apart. The heated portion will stretch to a long, narrow tube, optimally 1 mm in diameter, that can be cut to suitable lengths (approximately 9 cm) after cooling. One end of each cut length should be sealed by heating. Pulling too rapidly will produce capillaries too small in diameter, and pulling too rapidly with insufficiently softened glass will cause the capillary to break.

D. DECANTATION, FILTRATION, CENTRIFUGATION

Decantation. Sometimes adequate separation of a solid and a liquid can be achieved by decantation, especially if the solid is dense. Mix a solid and a liquid, and then allow the mixture to stand until the solid has settled; then carefully decant (pour off) the liquid, leaving the solid in the original container (Fig. D.1).

Gravity Filtration. The simplest method for separating a liquid and a solid is by gravity filtration through paper. A piece of filter paper is folded in

Figure D.1 Decanting a liquid from a solid.

half and then in half again, but not exactly at a right angle (Fig. D.2). Open it and tear off a small corner of the outside fold. The tear is made in the corner of the paper to allow a close seal to be made across the folded portion of the paper. Place the paper in a regular glass funnel, wetting the paper slightly to keep it in place. Support the funnel in a ring (or a ring plus a clay triangle) (Fig. D.3) with its stem touching the inside of a beaker. Pour the mixture into the funnel along a stirring rod, as in the figure. Do not allow the stirring rod to touch the paper. Never fill the filter paper more than two-thirds full. Finally, to complete the transfer, rinse the inside of the beaker and the stirring rod with at least two small portions of the solvent phase of the mixture contained in a wash bottle.

Filter paper is made in a variety of porosities. A fine filter paper will retain solids of very small particle sizes, but filtration is slow through such paper. When particles of a solid are large, a more porous paper is used, and filtration is faster. Occasionally, an analysis requires the filter paper carrying the solid to be burned. *Ashless* filter papers are available for this purpose.

In certain cases, when dealing with bulky precipitates or a need for rapid filtration, a *fluted* filter

Figure D.2 Steps in folding filter paper for use in filtering with a regular funnel. Notice that the second fold is not exactly a right angle.

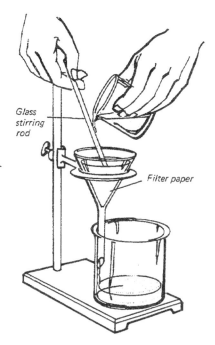

Figure D.3   Correct method of pouring the liquid to prevent spilling and splashing during gravity filtration.

Figure D.5   Safety trap for suction filtering when a water aspirator is used.

paper is desired. Such a paper, folded into many folds, presents a greater surface area to the liquid than a filter paper folded in the conventional way does.

Suction Filtration. When gravity filtration is too slow, suction filtration is often resorted to. A

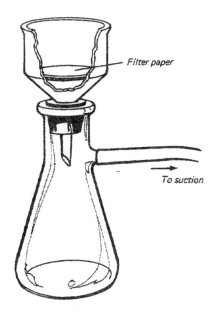

Figure D.4   Using the Büchner funnel for suction filtration.

Büchner funnel fitted with a rubber stopper is inserted into a suction flask (Fig. D.4). The sidearm of the flask is connected to a water aspirator or central vacuum source by a short piece of pressure tubing. A piece of filter paper slightly smaller than the funnel diameter is placed over the holes in the bottom of the funnel; the filter paper is first moistened with the solvent, and the aspirator turned on full. The aspirator may tend to cause water to back up in the flask. Thus, if the filtrate, that is, the liquid coming through the filter paper, is of value, a safety trap (Fig. D.5) should be connected between the flask and the aspirator. If the precipitate, namely, the solid retained on the filter paper, is to be washed, *stop the suction* and pour in the wash liquid; after a few minutes reapply the suction.

Centrifugation. A centrifuge substitutes centripetal force for gravity in the separation of solids from liquids (Fig. D.6). Whenever the centrifuge is used, it must be balanced or it may become damaged. Therefore, before centrifuging a mixture contained in a test tube or centrifuge tube, prepare another tube to balance it in the centrifuge by filling an identical tube with water until the liquid levels in both tubes are the same. Insert the tubes in opposite positions (at 180°) in the centrifuge, close the cover, and set the machine in motion. The time required for centrifugation depends on the particle size of the solid being separated; for example, crystalline solids require less time than colloidal precipitates. Allow the centrifuge to come to rest before removing the tubes. **Keep the cover closed and your hands away from the top of the centrifuge while it is rotating.**

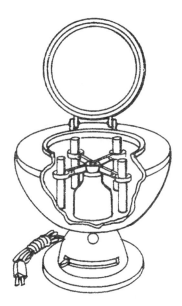

Figure D.6 Electric centrifuge for separating a precipitate from a liquid.

E. KEEPING SAMPLES DRY

Materials may absorb water if left exposed to the air. This is to be avoided, particularly if a sample of the material is to be weighed precisely. The desiccator is a device for preventing a substance from absorbing moisture (Fig. E.1). A simple device can be constructed from common laboratory equipment. Drierite is recommended as a drying agent since its effectiveness can be readily ascertained; it turns pink when it is no longer effective.

F. CONTROLLING FUMES

Obnoxious and poisonous fumes must not be allowed to escape into the laboratory. Experiments in which considerable fuming occurs must be carried out in fume hoods (see Fig. B.2.5, page 7) equipped with efficient exhaust fans.

Desiccant

Figure E.1 A laboratory desiccator.

G. MEASURING VOLUME

Volume of liquids can be estimated by medicine droppers, and beakers and flasks, but for more exact measurement, the following instruments are used: graduated cylinders, burets, volumetric (transfer) pipets, graduated (delivery) pipets, and volumetric flasks. To identify these instruments, see pages 12–13. The measuring instrument selected depends on the degree of accuracy needed for the particular determination. The instruments come in a variety of sizes. The capacity of each at a given temperature is stamped on by the manufacturer, together with etched lines for precise measure. For very precise work, all items of volumetric equipment must be calibrated.

Medicine Droppers. A very rough, but often satisfactory, method for estimating volume is by counting drops delivered from a medicine dropper 15 to 20 drops per milliliter, depending on the size of the tip and the surface tension of the liquid.

Beakers and Flasks. The volume stamped with the trademark on beakers and flasks is only approximately correct. Use this method for crude estimates of capacity only.

Volumetric Glassware. Volumetric equipment initially must be clean. Not only can dirt chemically contaminate an experiment, but its presence in a volume-measuring instrument may reduce precision, since dirty glassware does not drain properly. Special cleaning solutions are usually available in most laboratories, but using soap, water, and a brush—followed by thorough rinsing with water—is still an extremely effective method for cleaning glassware. The glassware is inverted to dry. Volumetric glassware should never be heated over a flame or dried with compressed air.

Graduated cylinders are perhaps the most available and versatile volume-measuring instruments in the laboratory. However, *burets* are constructed so that it is possible to measure (and deliver) volumes more accurately. For example, with a 50-mL buret, volume can be measured to within 0.01 mL, whereas with a 50-mL graduated cylinder, the value within which the volume can be measured is closer to 0.1 mL. The *volumetric pipet*, properly manipulated, can deliver volumes reliable to 1 part per thousand. The *graduated pipet* is not capable of such precision but is useful for delivering volumes for which volumetric pipets are not constructed, or perhaps not available in the laboratory. Some graduated pipets are calibrated "To Deliver," and are stamped "TD." The principal use of the *volumetric flask* is to prepare solutions of a specified concentration; that is, solutions containing a known amount of solute dissolved in a given volume of solution.

In pipetting a liquid, mouth suction must not be applied unless it is known with certainty that the liquid is harmless. Otherwise, a safety trap, a rubber aspirator bulb (see Fig. B.2.4, page 7), or a tube extension from the top of the pipet must be used. Keeping the pipet tip under the liquid surface and using suction (mouth or aspirator), you can draw the liquid above the graduation mark on the pipet. The forefinger is quickly applied to the end of the pipet, while the stem is held between the thumb and second finger (Fig. G.1), and the pipet removed from the liquid. Holding the pipet in a vertical position, one can adjust the liquid level by slowly releasing the pressure between the forefinger and the pipet—either by rotating the pipet with the thumb and second finger or by cautiously lifting the forefinger. The liquid is allowed to drain into an extra container until the bottom of the meniscus coincides with the desired graduation mark on the pipet. (The significance of the meniscus is discussed in the following paragraph.) Any drops adhering to the bottom of the tip should be removed. The measured volume is then delivered into the proper container. Pipets, on being emptied, are allowed to drain in a vertical position and held

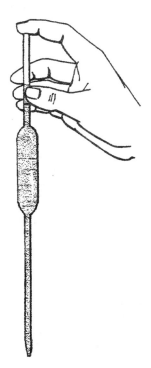

Figure G.1 Retaining the liquid in a pipet.

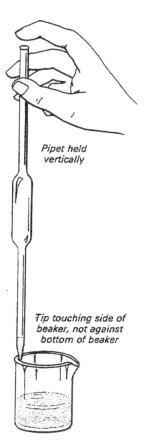

Pipet held vertically

Tip touching side of beaker, not against bottom of beaker

Figure G.2 Draining the liquid from a pipet.

for 20 seconds after draining; the tip is then touched to the wall of the receiver (Fig. G.2). The liquid left in the tip must not be removed; the calibration of the pipet allows for this volume.

The *meniscus* is the apparent downward curvature in the surface of a liquid contained in any narrow measuring tube, caused in part by surface tension. In pipets, and also in graduated cylinders or burets that are filled from the top, it is necessary to read the bottom of the meniscus with the eye horizontal to this surface (Fig. G.3). If the meniscus is not read at eye level, so that the front and rear parts of the graduation mark nearest the meniscus appear to coincide, parallax error in the reading will result. Determine this factor for yourself by comparing three readings of the same meniscus: one with the eye level horizontal, one with the eye directed from somewhat above, and one from somewhat below. Proper-lighting is necessary to see the meniscus clearly. One exception is that of the meniscus of mercury, which is read at the top. Mercury is the only metal that is a liquid, and is very dense.

Loss in volume in transfer of liquids can be an inherent error in experiments. When liquids are poured from cylinders (or other vessels), they can

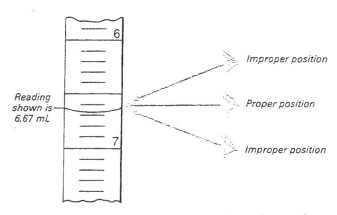

Reading shown is 6.67 mL

Improper position

Proper position

Improper position

Figure G.3 Close-up of a volumetric buret showing the meniscus. The correct reading is at the level of the bottom of the meniscus, read at eye level.

be prevented from splashing by being allowed to run down a stirring rod (see Fig. D.3, page 390).

Loss in volume of liquids being heated in test tubes due to "bumping" can be avoided if the test tube is less than half full and if suitable precautions are taken. Hold the test tube at an angle and heat at the bottom *with a small flame,* moving the tube slowly back and forth to prevent overheating at any one spot. Placing a boiling chip, such as a small piece of clay plate, in the liquid helps to prevent bumping. Solutions in test tubes are often more conveniently heated in water baths.

H. MEASURING MASS

One of the common operations in experimental chemistry is weighing. The following types of balances are found in the laboratory: the platform balance and the hanging-pan balance, the electronic top-loading balance, the single-pan analytical balance, and the electronic analytical balance. Specific instruction sheets are available for specific balance models.

Observe the following rules in any weighings:

1. Before you make any weighings, acquaint yourself with the type of balance you will use. Your instructor will demonstrate any specific features of your balance.
2. *Never* weigh chemicals directly on the balance pan; use a watch glass, beaker, or smooth paper as a container. Clean up any spills.
3. If there is an apparent difficulty in operating the balance, consult an instructor. Do not attempt any adjustments.
4. If you are using a hanging-pan balance, do not interchange pans; the pan is calibrated for its particular balance.

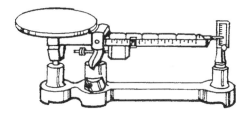

Figure H.1   Platform balance.

5. Do not remove analytical balances from their location.
6. If you are using a single-pan analytical balance, arrest the balance beam before removing the object from the pan. Do not jar the mechanism.
7. Weigh volatile or corrosive samples in stoppered containers.
8. Do not weigh objects that are not at room temperature.

Check your ability to use a balance in the following manner. Select two small, clean, dry objects, such as a watch glass and a crucible lid. Weigh the two objects separately, record each weight, and add the results. Then weigh the two objects together. Determine whether your results check. If not, try again until you are sure you can weigh accurately.

The Platform Balance (Fig. H.1). The platform balance is a semiquantitative type weighing to 0.1 g. It is used primarily for weighing large amounts of materials (over 100 g).

The Hanging-Pan Balance (Fig. H.2). The hanging-pan balance weighing to 0.01 g is both convenient and rugged. It is usually used for amounts less than 100 g, although some balances of this type have a greater capacity.

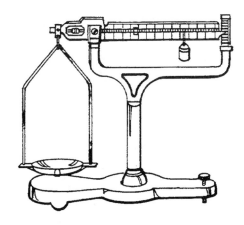

Figure H.2   Hanging-pan balance.

Figure H.3   Electronic top-loading balance.

The Electronic Top-Loading Balance (Fig. H.3). Some models of top-loading balances offer more precision than hanging-pan balances, though less than analytical balances. Such top-loading balances can weigh to 0.001 g. This type is the most desirable—in terms of time-saving ease of operation. Some models have one control bar; others may have two buttons: on/tare and off. The sample is placed on the pan, and the weight is easily read from a liquid crystal display.

Some laboratories may have digital top-loading balances, which are less convenient to use than electronic models because they require more steps in operation.

The Single-Pan Analytical Balance (Fig. H.4). The single-pan model can allow a precision of 0.0001 g (0.1 mg). All weighing is done by dialing. The main precaution in this type of balance is to minimize vibration. Detailed procedure depends on the particular make of balance used.

The Electronic Analytical Balance (Fig. H.5). The usual 100-g capacity balance of this type has

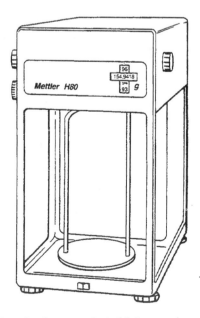

Figure H.4   Single-pan analytical balance.

Figure H.5   Electronic analytical balance.

0.0001 g (0.1 mg) readability, with a liquid crystal display. It is rugged and reliable, with uncomplicated operation.

Most of the exercises in this manual require a balance that can weigh to 0.01 g. Certain exercises benefit from a balance that can weigh to 0.001 g.

I. MEASURING TEMPERATURE; CALIBRATING A THERMOMETER

The markings on thermometers are occasionally found to be in error. Calibrating a thermometer involves determining the position of the mercury thread at the temperature of a mixture of crushed ice and water (defined as 0°C), and the temperature of boiling water at 760 mm Hg of pressure (defined as 100°C).

At 0°C. Suspend a thermometer from a clamp in such a position that its lower 7 cm are immersed in a mixture of crushed ice and water. The bulb of the thermometer must not touch the sides or bottom of the vessel. Read the thermometer, estimating at the nearest 0.1°C, after 5 minutes of immersion.

At 100°C. Locate the barometer in your laboratory. Assemble an apparatus as shown in Fig. I.1. Place 5 mL of distilled water in the test tube. Place the thermometer bulb one cm above the water level, and add one small boiling chip. Slowly heat the water to a smooth boil, to avoid splashing. While

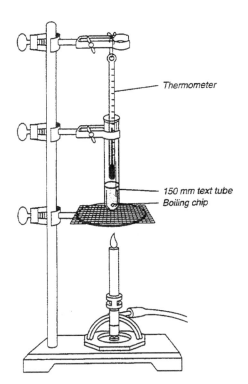

Figure I.1 Apparatus for determining the boiling point of water.

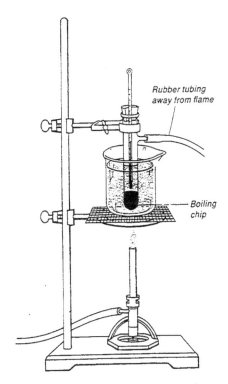

Figure J.1 Apparatus for determining the boiling point of a substance.

the water is still boiling, read the thermometer to the nearest 0.1°C. Read the barometer.

From the results obtained, record the corrections to be applied at 0°C and at 100°C, together with the number, if any, of the thermometer calibrated. For the true boiling point of water at a given pressure, refer to the table below.

Barometric pressure (mm Hg)	Boiling point of water (°C)
745	99.4
750	99.6
755	99.8
760	100.0

From your data, a graph can be constructed that shows the actual temperatures when the thermometer is read in the intervals between 0°C and 100°C. (These corrections are assumed to be of equal reliability.)

J. DETERMINING BOILING POINT

If the liquid to be tested is *known* to be *nonflammable*, See Fig. I.1. Otherwise, follow the directions below.

Obtain a sidearm test tube. (If a sidearm test tube is not available, use a 200 × 25-mm Pyrex test tube equipped with a two-hole stopper carrying a short glass bend in one of the holes.) Fill the test tube to a depth of about 3 cm with the liquid of which the boiling point is to be determined. Add one small boiling chip to the liquid to ensure even boiling. Assemble the apparatus shown in Fig. J.1. Suspend the bulb of the thermometer 1 cm above the surface of the liquid. Immerse the clamped test tube in the 600-mL beaker of water as shown. Lead an appropriate length of rubber tubing connected to the sidearm of the test tube (or to the glass bend, if the alternative setup is used), away from burner flames. Heat the water gradually and watch for changes in the liquid in the test tube. Record the temperature at which the liquid in the test tube boils freely. The temperature should reach a stable point; record this temperature as the boiling point.

If the liquid to be tested has a boiling point higher than 100°C, a beaker containing oil rather than water must be used.

K. DETERMINING MELTING POINT

Push the open end of a capillary tube into a small quantity of the substance of which the melting point is to be determined. If capillary tubes are not supplied, they may be made (see Part 3C, page 388). Tap the tube on the desk to pack the crystals to a depth of about 2 mm. Attach the capillary tube to a thermometer by means of a rubber band or suitable section of rubber tubing (see Fig. K.1), and assemble the setup as shown. Lower the thermometer so that the *lower* end of the capillary tube is below the surface of the bath (a 250-mL beaker containing water or oil), but do not allow any bath liquid to enter the capillary tube or be in contact with the rubber band. Heat the bath slowly (2–4°C per minute). Note the temperature at which the substance has just completely melted; that is, when the last remaining crystal has just disappeared.

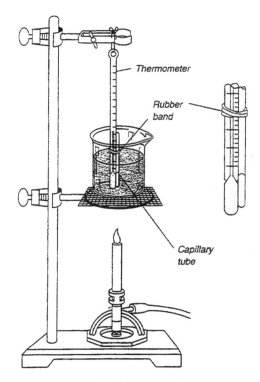

Figure K.1 Apparatus for determining the melting point of a substance.

Part 4

Semimicro Qualitative Analysis

General Laboratory Directions

INTRODUCTION

Qualitative analysis has two principal objectives. One is the practical goal of learning careful laboratory manipulation, critical observation, and logical interpretation of observed results. The other is the understanding of reasons for the analytical procedures and results in terms of the theory of ionic equilibria, especially that relating to weak electrolytes, solubility products, complex ions, and oxidation–reduction.*

Semimicro qualitative analysis is a method of analysis employing techniques whereby the reactions and procedures used in macro work may be reliably carried out on a reduced scale. Analysis on the macro scale is made on volumes of solutions of the order of 10 to 100 mL and with ordinary test tubes, beakers, and funnels. In semimicro analysis, volumes of solutions from 1 drop to about 5 mL are employed, with small test tubes, centrifuge tubes, capillary syringes, and medicine droppers used to carry out the separations and identification tests. This leads to a striking reduction in the consumption of reagents. Usually the analyses can also be carried out more rapidly by semimicro methods.

Even though rather definite directions are given for the analysis to be carried out, no two analyses will be exactly alike. For this reason directions should never be followed blindly, but with careful thought, and procedures should be adapted to the particular problem at hand.

---

*Part 4 primarily presents the laboratory procedures. The same procedures are given in *General Chemistry with Qualitative Analysis*, 10th ed., by Robinson et al. (Houghton Mifflin Co., Boston, Mass.: 1997), which also discusses the principles involved in the analytical procedures.

Equipment

The Appendix lists the apparatus and solid reagents and solutions that each student will need to carry out the laboratory work of this part of the course. Obtain from the stockroom any apparatus that may be required but is not in the desk. Wash all of the apparatus that will be employed in the analyses before beginning your laboratory work.

Laboratory Assignments

Below is a suggested set of laboratory assignments. The number of unknowns required will be specified by the instructor.

1. Construct a wash bottle, stirring rods, and capillary syringes.
2. Analyze a known solution containing all the cations of Group I.
3. Analyze an unknown solution based on the Group I cations.
4. Analyze a known solution containing all the cations of Group II.
5. Analyze an unknown solution based on the Group II cations.
6. Analyze a known solution containing all the cations of Group III.
7. Analyze an unknown solution based on the Group III cations.
8. Analyze a known solution containing all the cations of Group IV.

9. Analyze a known solution containing all the cations of Group V.
10. Analyze an unknown solution based on the cations of Groups IV and V.
11. Analyze a general unknown solution based on all the cations of all the groups.
12. Analyze a known salt mixture containing the oxidizing anions (but no reducing anions).
13. Analyze an unknown salt mixture based on the oxidizing anions.
14. Analyze a known salt mixture containing the reducing anions (but no oxidizing anions).
15. Analyze an unknown salt mixture based on the reducing anions.
16. Analyze a salt mixture for both cations and anions.
17. Analyze an alloy.

Wash Bottle

The *plastic* wash bottle (see pages 12–13, Common Laboratory Equipment) is currently in use in many laboratories. If you are not supplied with one, a satisfactory bottle can be constructed using a 250-mL Florence flask and 6-mm glass tubing, as shown in Fig. S.1. Ask you instructor for directions. Keep the wash bottle filled with *distilled water* for use in the analytical procedures. Ordinary tap water contains such ions as Ca^{2+}, Mg^{2+}, Fe^{3+}, Al^{3+}, Cl^-, SO_4^{2-}, and HCO_3^-. Since these ions are among those to be tested for in the unknown solutions, **use only distilled water** in the procedures and in the cleaning of apparatus. **Whenever water is mentioned in the following procedures, distilled water is required.**

Stirring Rods

Make or obtain at least five glass stirring rods, approximately 15 cm in length and 3 mm in diameter. The ends of each rod must be fire-polished.

Capillary Syringes

Standard medicine droppers of approximately 1-mL capacity are used for adding solutions of reagents. These droppers deliver about 20 drops per milliliter. A second type of dropper, called a capillary syringe (Pasteur pipet), is needed for the removal of liquids from precipitates held in small test tubes or centrifuge tubes. If not available, the capillary syringes may be made from glass tubing as follows (More information on heating with a Bunsen burner, and on glassworking, is found in Parts 3B and 3C.): Heat the middle portion of a 17-cm piece of 8-mm glass tubing over a Bunsen flame with rotation until the glass softens. Remove the tube from the flame and slowly draw it out until the bore is about 1 mm. When the tube has cooled, cut the capillary at the midpoint and fire-polish the capillary ends. Flare the wide ends of the tubes by heating until soft and quickly pressing down against a flat metal or other flat nonflammable surface. When cold, attach medicine-dropper bulbs to the flared ends of the syringes. These syringes will deliver approximately 40 drops per milliliter.

Reagents

The solids and solutions called for in the analytical procedures will be stored in the 10-mL reagent bottles (Fig. S.2) to be found in your desk. Fill these reagent bottles with the chemicals required in the analysis of each group prior to starting the analysis of that group. See the Appendix for a list of these reagents. Only a small quantity of the starred reagents (*) will be needed during the course, so fill the bottles only about one-fourth full of these reagents.

Before filling a reagent bottle make sure that it is perfectly clean. To avoid mistakes, it is a good idea to label each bottle before filling it.

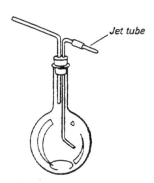

Figure S.1 A laboratory-constructed wash bottle.

Figure S.2 Reagent bottle equipped with dropper used in semimicro analytical work.

Precipitation

Practically all of the precipitations are carried out in either 4-mL Pyrex test tubes or 2-mL conical test tubes. Check for completeness of precipitation by adding a drop of reagent to the solution (centrifugate) obtained in the separation of the precipitate. If the addition of more reagent to the solution shows that precipitation is incomplete, separate the mixture and test the second solution for completeness of precipitation.

The precipitating agent should be added slowly, preferably from a medicine dropper, with vigorous shaking or stirring of the reaction mixture. The formation of larger crystals of the precipitate is favored by warming the solution, and separation of the precipitate should not be attempted before the crystals become large enough to settle.

A slight excess of the precipitating agent is added to reduce the solubility of the precipitate by the common ion effect. On the other hand, a very large excess of the precipitating agent should be avoided since it may actually increase the solubility of the precipitate. For example, in precipitating silver chloride a large excess of Cl^- will bring about the formation of $AgCl_2^-$ and thereby increase the solubility of AgCl. Many precipitates are dissolved, at least partially, by the formation of complexes of this type.

Centrifugation of Precipitates

A precipitate may be separated from a liquid by means of a centrifuge (see Fig. S.3). By rotating a mixture of solid and liquid at high speed in a centrifuge, the more dense precipitate is forced to the bottom of the containing tube by a centrifugal force that is many times the force of gravity. This accounts for the much shorter time required for settling of the precipitate when centrifugation is employed. Colloidal precipitates require longer centrifugation than do crystalline precipitates due to the small size of the colloidal particles.

A centrifuge may be damaged if allowed to run unbalanced. Therefore, before centrifuging a precipitate contained in a test tube or centrifuge tube, prepare another tube to balance it in the centrifuge by filling an identical tube with water until the liquid levels in both tubes are the same. Insert the tubes in opposite positions in the centrifuge and set the machine in motion. Allow the centrifugation to continue for 30 seconds. After the machine has come to rest, remove the tubes.

Transfer of the Centrifugate

After centrifugation the precipitate will be found packed in the bottom of the tube. The supernatant liquid or centrifugate is separated from the precipitate by holding the tube at an angle of about 30 degrees (see Fig. S.4) and removing the supernatant liquid by slowly drawing it into a capillary syringe. The tip of the syringe is held just below the surface of the liquid; as the pressure on the bulb is slowly released, causing the liquid to rise in the syringe, the capillary is lowered into the tube until all of the liquid is removed. As the capillary approaches the bottom of the tube, the tip must not be allowed to stir up the mixture by touching the precipitate.

Washing of the Precipitate

The precipitate left in the tube after the removal of a supernatant liquid is still wet with a solution containing the ions of this liquid. The precipitate is washed, usually with water, to dilute the solution adhering to the precipitate. The wash liquid is

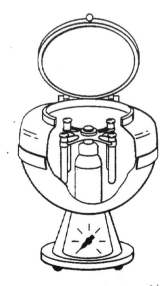

Figure S.3 Electric centrifuge. Cutaway shows arrangement for holding test tubes used in centrifuging a precipitate.

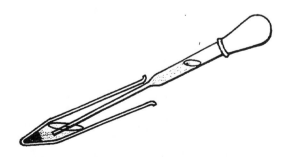

Figure S.4 Using a capillary syringe for removing a supernatant liquid after centrifuging.

added to the precipitate and the mixture is stirred thoroughly. After centrifugation, the washings are removed by a capillary syringe as previously described (see Fig. S.4). Usually a precipitate is washed at least twice. The first wash liquid is usually saved and added to the first filtrate. If the precipitate must be transferred to another container, the reagent to be used is added, the mixture is stirred well, and then it is poured into the other container. After the precipitate has settled, the supernatant liquid may be employed to remove any precipitate remaining in the centrifuge tube.

Dissolution and Extraction of Precipitates

When all or a part of a precipitate is to be brought into solution by a reagent, the solvent is added to the precipitate that is in the centrifuge tube and the mixture is stirred. The mixture is then separated by centrifugation and the operation is repeated using fresh solvent. Oftentimes the extraction of a precipitate is more efficient at an elevated temperature.

Heating of Mixtures of Solutions

Whenever it is necessary to heat a mixture for the purpose of bringing about a precipitation or for dissolving or extracting a precipitate, the test tube or centrifuge tube is placed in a water bath (see Fig. S.5) maintained at a suitable temperature. It will be found convenient to keep the water hot in the water bath throughout the work period.

Evaporation

It is often necessary to heat solutions to boiling and to hold them at the boiling temperature in order to concentrate them, or to remove volatile acids or bases, or even to evaporate a solution to dryness. Evaporation should be carried out in a small casserole or porcelain evaporating dish. Agitate the contents of the casserole constantly while the heating continues. Avoid evaporation of solutions contained in small test tubes because the contents of the tube may be lost due to overheating.

Cleaning Glassware

Because small amounts of contaminants may give rise to erroneous results, all glassware used in the analytical procedures should be thoroughly cleaned before it is used. The cleaning should be done with a brush and some cleansing powder such as a synthetic detergent. The apparatus should then be rinsed first with tap water and finally with distilled water. Test-tube brushes and centrifuge tube brushes are available. Medicine droppers, capillary syringes, and stirring rods should be cleaned, rinsed, and stored in a beaker of distilled water.

Flame Tests

Flame tests are made using a platinum wire sealed in the end of a piece of glass tubing. The wire is first looped at the end and then cleaned by dipping the looped end in 6 M HCl and then heating it in the Bunsen flame (see Fig. S.6). Rather than us-

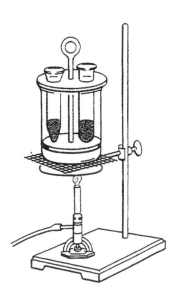

Figure S.5 Individual water bath for heating reaction mixtures.

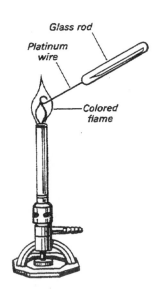

Figure S.6 Platinum wire mounted in glass rod used in making flame tests for unknown ions.

ing the hottest portion of the flame, the looped end of the wire should be brought slowly up to the edge of the flame. The platinum wire should not be held in the reducing part of the flame. The operation of dipping in acid and heating in the flame should be repeated until the wire no longer imparts a color to the flame. The wire loop is dipped in the solution to be tested and then heated in the flame. The memory should not be relied upon in judging the color imparted to the flame. Instead a known solution should be tested and the color compared to that given by the unknown.

Known Solutions

In order to acquaint you with the details of the analytical procedures before any attempt to analyze a solution of unknown composition, known solutions containing all the ions of a given group are provided. This is done because it is necessary that students become familiar with the separations and confirmatory tests for the ions of the group by practicing on a known solution before trying to determine the ions present in an unknown solution. Note the quantities of precipitates obtained and the colors of precipitates and solutions as the analysis of the known solution is carried out. These observations and the equations for the reactions involved should be recorded.

Unknown Solutions

When the analysis of a known solution has been satisfactorily completed, you should at once pre-pare for a possible oral quiz by the instructor on the details of the analysis of the group. This quiz may include questions on the reactions involved, the theory of the separations and confirmatory tests, and the colors of precipitates, ions, and solutions. After the instructor's questions have been satisfactorily answered, or if a quiz is not required, a sample of an unknown may be obtained for analysis. The unknown may contain any or all of the ions of the group or groups being studied.

Unknown Report

All observations should be recorded immediately. The Unknown Report form (see the Appendix of this book) may be used for this purpose. Equations should be written to indicate the behavior of each ion with the reagents with which it comes in contact.

In reporting the results of the analysis of an unknown, an Unknown Report form should be filled out in a manner similar to the sample one on page 405 and presented to your instructor for grading. Report forms are found in the Appendix of this book. The answers to the Questions and Problems at the end of each analysis should be written on a separate sheet of paper and attached to the completed Unknown Report form.

NAME _____ SECTION _____ DATE _____

INSTRUCTOR'S APPROVAL _____ IONS FOUND _____ GRADE _____

CATION UNKNOWN REPORT (SAMPLE)

No.	Substance	Reagent	Result	Inference or conclusion	Precipitate or residue	Centrifugate or solution
1	Group I	HCl	White ppt	Group I present	One or more of AgCl, PbCl$_2$, Hg$_2$Cl$_2$	Possibly Pb^{2+}
2	Ppt from No. 1	Hot water	No visible action	Hg$_2^{2+}$ and/or Ag$^+$ present; Pb^{2+} uncertain	Hg$_2$Cl$_2$ and/or AgCl	Possibly Pb^{2+}
3	Solution from No. 2 or 1	CH$_3$CO$_2$H K$_2$CrO$_4$	Yellow ppt	Pb^{2+} present	PbCrO$_4$	
4	Residue from No. 2	NH$_3$ + H$_2$O	Residue dissolves completely	Hg$_2^{2+}$ absent; Ag$^+$ probable		Ag(NH$_3$)$_2^+$
5	Filtrate from No. 4	HNO$_3$	White ppt	Ag$^+$ present	AgCl	
6						

Analysis for Cations

INTRODUCTION

The metals may be classified into groups in two distinctly different ways: (1) according to the Periodic Table groupings, which reflect similarities and differences in electronic structures and gradations in physical and chemical properties, and (2) according to the way in which the metals are associated in their systematic identification, or qualitative analysis. The latter classification depends upon the solubilities of the various salts formed by the metals. It is of value to consider both groupings for each metal studied.

Qualitative analysis pertains to the identification of the constituents present in a sample of a substance, a mixture of substances, or a solution. In the qualitative analysis of a solution that may contain any or all of the common metal ions, the first step is that of separating the ions of the metals into several groups, each of which contains ions exhibiting a common chemical property that is the basis of the separation.

The Metals of Analytical Group I. When dilute hydrochloric acid is added to a solution containing all of the common metal ions (and ammonium, NH_4^+), the chlorides of mercury(I), silver, and lead precipitate. The chlorides of all the other common metal ions are soluble and can be separated from those of Group I by filtration or centrifugation.

The Metals of Analytical Group II. After the Group I chlorides have been separated, the solution is made 0.3 M in hydrochloric acid, and the Group II metals are precipitated as sulfides upon the generation of hydrogen sulfide in the solution. The precipitate formed consists of the sulfides of lead, bismuth, copper, cadmium, mercury(II), arsenic, antimony, and tin.

The Metals of Analytical Group III. After the Group II sulfides have been separated, the solution is saturated with hydrogen sulfide and then an excess of aqueous ammonia is added to the solution. Under these conditions the sulfides of cobalt, nickel, manganese, iron, and zinc, and the hydroxides of aluminum and chromium are precipitated.

The Metals of Analytical Group IV. The Group IV metals (barium, strontium, and calcium) are precipitated as the carbonates from the centrifugate of Group III by ammonium carbonate in the presence of aqueous ammonia and ammonium chloride.

The Metals of Analytical Group V. The centrifugate from Group IV contains the ions of sodium, potassium, magnesium, and ammonium, which constitute Group V.

It is important to remember that, in general, a precipitating agent for one group will also precipitate all previous groups.

The general flow sheet for the cation group separations is shown in Table S.1.

TABLE S.1. FLOW SHEET OF GROUP SEPARATIONS (FORMULAS FOR PRECIPITATES ARE UNDERLINED)

	$\underline{Hg_2Cl_2}$ / $\underline{AgCl}$ / $\underline{PbCl_2}$	**Group I Chlorides**					
Hg_2^{2+}							
Ag^+							
Pb^{2+}	Pb^{2+}	$\underline{PbS}$	**Group II Sulfides**				
Bi^{3+}	Bi^{3+}	$\underline{Bi_2S_3}$					
Cu^{2+}	Cu^{2+}	$\underline{CuS}$					
Cd^{2+}	Cd^{2+}	$\underline{CdS}$					
Hg^{2+}	Hg^{2+}	$\underline{HgS}$					
As^{3+}	As^{3+}	$\underline{As_2S_3}$					
Sb^{3+}	Sb^{3+}	$\underline{Sb_2S_3}$					
Sn^{4+}	Sn^{4+}	$\underline{SnS_2}$					
Co^{2+}	Co^{2+}	Co^{2+}	$\underline{CoS}$	**Group III Sulfides and Hydroxides**			
Ni^{2+}	Ni^{2+}	Ni^{2+}	$\underline{NiS}$				
Mn^{2+}	Mn^{2+}	Mn^{2+}	$\underline{MnS}$				
Fe^{3+}	Fe^{3+}	Fe^{2+}	$\underline{FeS}$				
Al^{3+}	Al^{3+}	Al^{3+}	$\underline{Al(OH)_3}$				
Cr^{3+}	Cr^{3+}	Cr^{3+}	$\underline{Cr(OH)_3}$				
Zn^{2+}	Zn^{2+}	Zn^{2+}	$\underline{ZnS}$				
Ba^{2+}	Ba^{2+}	Ba^{2+}	Ba^{2+}	$\underline{BaCO_3}$	**Group IV Carbonates**		
Sr^{2+}	Sr^{2+}	Sr^{2+}	Sr^{2+}	$\underline{SrCO_3}$			
Ca^{2+}	Ca^{2+}	Ca^{2+}	Ca^{2+}	$\underline{CaCO_3}$			
Mg^{2+}	Mg^{2+}	Mg^{2+}	Mg^{2+}	Mg^{2+}	**Group V Soluble Ions**		
NH_4^+	NH_4^+	NH_4^+	NH_4^+	NH_4^+			
Na^+	Na^+	Na^+	Na^+	Na^+			
K^+	K^+	K^+	K^+	K^+			

Arrows: $\xrightarrow{HCl}$; $\xrightarrow[H_2S]{0.3\ M\ HCl}$; $\xrightarrow[H_2S]{NH_4Cl,\ NH_3 + H_2O}$; $\xrightarrow[(NH_4)_2CO_3]{NH_3 + H_2O,\ NH_4Cl}$

The Analysis of Group I

Begin with these procedures for the analysis of a solution of a known or unknown containing the metal ions. Depending upon the nature of the known or unknown, the solution may contain only the ions of Group I or it may contain the ions of one or all of the Groups I–V. If your sample is a general unknown, reserve a part of the original solution to test for the ammonium ion in Group V.

Procedure 1. *Precipitation of Group I:* Hg_2^{2+}, Ag^+, Pb^{2+}. To 10 drops of the solution to be analyzed add enough water to make a total volume of 1 mL. Add 2 drops of 6 M HCl. Avoid using excess HCl since the soluble complex ions $[AgCl_2]^-$ or $[PbCl_2]^{2-}$ may form. Centrifuge and separate the solution (centrifugate) from the precipitate. If your sample is a general unknown, reserve the solution (centrifugate) for the analysis of Groups II–V. If your sample is a Group I known or unknown, test the solution for lead as directed in Procedure 2. Treat the precipitate according to Procedure 2.

Procedure 2. *Precipitate from Procedure 1:* Hg_2Cl_2, $AgCl$, $PbCl_2$. Extract the precipitate twice with 10 drops of boiling hot water to dissolve the $PbCl_2$. Reserve the residue for Procedure 3. Add 2 drops of 1 M K_2CrO_4 to the solution (hot water extract). A yellow precipitate, $PbCrO_4$, confirms the presence of lead.

Procedure 3. *Residue from Procedure 2:* Hg_2Cl_2 and $AgCl$. Extract the residue twice with 5 drops of 4 M aqueous ammonia. A black or gray residue confirms the presence of mercury(I) ion. Reserve the aqueous ammonia extract for Procedure 4.

Procedure 4. *Solution from Procedure 3:* $Ag(NH_3)_2Cl$. Add 4 M HNO_3, with stirring, until the solution is acid to litmus. A white precipitate (or cloudiness) of $AgCl$ confirms the presence of silver.

The flow sheet for Group I analysis is shown in Table S.2.

TABLE S.2. GROUP I FLOW SHEET

$$Hg_2^{2+} \\ Ag^+ \\ Pb^{2+}$$ $\Big\}$ $\xrightarrow{HCl}$ $\begin{array}{l} Hg_2Cl_2 \\ AgCl \\ PbCl_2 \end{array}$ $\Big\}$ $\xrightarrow{Hot\ H_2O}$ $\begin{array}{l} Hg_2Cl_2 \\ AgCl \end{array}$ $\Big\}$ $\xrightarrow{NH_3\ +\ H_2O}$ $\begin{array}{l} Hg\ (black)\ +\ HgNH_2Cl\ (white) \\ Ag(NH_3)_2^+ \xrightarrow[Cl^-]{HNO_3} AgCl\ (white) \end{array}$

$Pb^{2+} \xrightarrow{CrO_4^{2-}} PbCrO_4$ (yellow)

$Pb^{2+} \xrightarrow{CrO_4^{2-}} PbCrO_4$ (yellow)

Questions and Problems for Group I

1. Why must a large excess of chloride ion be avoided in the precipitation of the Group I chlorides?

2. In the case of an unknown containing only the cations of Group I, why is it advisable to make a confirmatory test for lead on the filtrate from the Group I precipitation?

3. What general statement can be made concerning the solubility of common chloride salts other than those of the analytical Group I cations?

4. Give the color of each of the following: $AgCl$, $PbCl_2$, Hg_2Cl_2, $PbCrO_4$, Hg, $HgNH_2Cl$, and $Ag(NH_3)_2\ Cl$ in solution.

5. In terms of ionic equilibria and solubility product theory, explain the dissolution of silver chloride in aqueous ammonia and its reprecipitation with nitric acid.

6. Select a reagent used in the analysis of Group I that will in one step separate each of the following pairs: (a) $AgCl$, $CuCl_2$, (b) $AgCl$, $PbCl_2$, (c) Hg_2Cl_2, $PbCl_2$, (d) Hg_2Cl_2, $AgCl$, (e) Hg_2^{2+}, Hg^{2+}.

7. Why don't the slightly soluble oxychlorides of bismuth and antimony precipitate with the Group I chlorides?

The Analysis of Group II

The solution to be analyzed may be a Group II known or unknown, or it may be the solution from the Group I separation. In either case proceed according to the following procedure.

Procedure 1. *Precipitation of Group II:* Pb^{2+}, Bi^{3+}, Cu^{2+}, Cd^{2+}, Hg^{2+}, As^{3+}, Sb^{3+}, Sn^{2+}, Sn^{4+}. Add 5 drops of 4 M HNO_3 to 10 drops of the known or unknown (or to the centrifugate from

the Group I separation) and evaporate the solution over a microflame to a moist solid residue. Cool and add 10 drops of water. Add 1 M aqueous ammonia until the solution (or mixture) is just basic to litmus. (Dip a stirring rod into the solution and then touch the moist end of the rod to a piece of red and blue litmus paper in testing for acidity or basicity of solutions.) Add 1.0 M HCl until the solution is just acidic to litmus. Now add 2 drops of 6 M HCl and dilute the solution to 1.5 mL. Judge the volume by comparing with a measured amount in a similar tube. Add 10 drops of 5% thioacetamide solution. **Caution: Avoid skin contact with this solution, and wash hands thoroughly on leaving the laboratory.**

The hydrolysis of thioacetamide, CH_3CSNH_2, conveniently provides an aqueous solution of hydrogen sulfide:

$$CH_3CSNH_2(l) + 2H_2O(l) \longrightarrow \\ CH_3CO_2(aq)^- + NH_4(aq)^+ + H_2S(aq)$$

The following equilibria provide the S^{2-} ion to combine with the cations:

$$H_2S \rightleftharpoons HS^- + H^+$$
$$HS^- \rightleftharpoons H^+ + S^{2-}$$

The solution should now be 0.3 M with respect to the hydrogen ion if the directions have been followed carefully. Heat this solution contained in a test tube in a hot water bath for at least 5 minutes. Add 1 mL of water and heat the mixture for another 10 minutes. Centrifuge and separate the precipitate; reserve the centrifugate for the analysis of Groups III–V in the case of a general unknown. Wash the precipitate with 10 drops of 0.1 M HCl twice. Add the first 10 drops of wash solution to the original centrifugate and discard the rest. Be certain that you are discarding the correct portion of the wash solution.

Procedure 2. *Precipitate from Procedure 1:* PbS, Bi_2S_3, CuS, CdS, HgS, As_2S_3, Sb_2S_3, SnS_2. Add a mixture of 12 drops of 4 M NaOH and 4 drops of thioacetamide solution to the precipitate. Heat the mixture in a hot water bath for 5 minutes, stirring occasionally. Separate the mixture by centrifugation and reserve the solution (thio salts of the Division B ions) for Procedure 6. The residue consists of the sulfides of the ions of Division A.

Procedure 3. *Residue from Procedure 2:* PbS, CuS, CdS, Bi_2S_3. Wash the residue with 1 mL of water to which 2 drops of 1 M NH_4NO_3 have been added. Discard the wash solution. Be certain of the nature of the discarded solution. Add 15 drops of 6 M HNO_3 to the residue, heat with stirring in a hot water bath for several minutes to dissolve the sulfides, the yellow substance floating on top of the solution. Separate and discard any sulfur that may have been formed by the nitric acid oxidation of sulfide ions and transfer the solution to a casserole or evaporating dish. Add 5 drops of 4 M H_2SO_4, **carefully and gently** evaporate to white SO_3 fumes but with residue still moist (*very important*). Cool and add 1 mL of water. Warm the mixture and stir the precipitate. Separate the mixture. Reserve the solution for Procedure 4. Wash the residue, $PbSO_4$, with a few drops of water and discard the wash solution. Extract the residue with a mixture of 5 drops of 1 M ammonium acetate, $NH_4CH_3CO_2$, and 1 drop of 1 M acetic acid, CH_3CO_2H. Add 1 drop of 1 M K_2CrO_4 to the extract. Scratch the inside wall of the test tube with a glass rod to initiate precipitation. The formation of a yellow precipitate, $PbCrO_4$, confirms the presence of lead.

Procedure 4. *Solution from Procedure 3:* Bi^{3+}, Cu^{2+}, Cd^{2+}. Add 15 M aqueous ammonia dropwise to the solution until it is distinctly basic to red and blue litmus paper (about 5 drops). The development of a deep blue color in the solution indicates the presence of the tetraamminecopper(II) ion, $Cu(NH_3)_4^{2+}$. The formation of a white precipitate indicates the presence of Bi^{3+} as $Bi(OH)_3$. Separate the mixture and reserve the solution for Procedure 5. Add several drops of freshly prepared sodium stannite solution* to the precipitate. The immediate formation of a black residue (finely divided bismuth) confirms the presence of Bi^{3+}.

---

* Preparation of sodium stannite: In a separate test tube place 3 drops of 0.4 M $SnCl_2$ and add *sufficient* 4 M NaOH to dissolve the white precipitate that first forms.

Procedure 5. *Solution from Procedure 4:* $Cu(NH_3)_4^{2+}$ (blue), $Cd(NH_3)_4^{2+}$ (colorless). If the solution is colorless, a trace of copper still may be present. If colorless, place 10 drops of the solution from Procedure 4 in a test tube (save the rest), acidify (test with litmus paper) with acetic acid, CH_3CO_2H, and then add several drops (5 or 6) of 0.1 M $K_4Fe(CN)_6$. The formation of a pink precipitate, $Cu_2Fe(CN)_6$, indicates the presence of a low concentration of Cu^{2+}; the formation of a white precipitate indicates the absence of Cu^{2+} and the probable presence of Cd^{2+} as $Cd_2Fe(CN)_6$.

If copper(II) ions are present, take *another* 10 drops of the solution from Procedure 4, in a clean test tube, add 1.0 M KCN (obtained from the instructor) until the solution is colorless and then add 2 drops of thioacetamide to the solution. Heat the mixture in a hot water bath. The formation of a yellow or olive-green precipitate (CdS) confirms the presence of cadmium.

If copper(II) ions are absent, test for Cd^{2+} as described above but leave out the KCN.

Procedure 6. *Solution from Procedure 2:* Thio (Sulfur) Salts of the Division B Ions, HgS_2^{2-}, AsS_3^{3-}, SbS_3^{3-}, SnS_3^{2-}. Add 1 M HCl until the solution is just acid to litmus. Heat the mixture in a hot water bath for several minutes. Separate the mixture and discard the solution. The precipitate consists of HgS, As_2S_3, Sb_2S_3, and SnS_2.

Procedure 7. *Precipitate from Procedure 6:* HgS, As_2S_3, Sb_2S_3, SnS_2. Add 1 mL of 6 M HCl to the precipitate and stir the mixture. Heat the mixture in a hot water bath and then separate the residue, *saving the centrifugate*. Add 15 drops of 6 M HCl to the residue and heat the mixture. Separate the residue and reserve the combined centrifugates for Procedure 11. Treat the residue as in Procedure 8.

Procedure 8. *Residue from Procedure 7:* HgS and As_2S_3. Add 12 drops of 4 M aqueous ammonia and 6 drops of 3% hydrogen peroxide to the residue. Stir the mixture and heat it in a hot water bath for several minutes. Separate the residue (HgS) and reserve the solution (AsO_4^{3-}) for Procedure 10.

Procedure 9. *Residue from Procedure 8:* HgS. To the black residue add 6 drops of 5% NaClO and 2 drops of 6 M HCl. Stir the mixture, add 1 mL of water, and separate the sulfur from the solution. Heat the solution to boiling. Add 2 drops of 1 M $SnCl_2$ to the solution. The formation of a white, gray, or black precipitate confirms the presence of mercury.

Procedure 10. *Solution from Procedure 8:* AsO_4^{3-}. Add 2 drops of 15 M aqueous ammonia and 5 drops of magnesia mixture to the solution. The formation of a white precipitate, $MgNH_4AsO_4$, frequently slow in forming, indicates the presence of arsenate ions.

Procedure 11. *Solution from Procedure 7:* $SnCl_6^{2-}$ and $SbCl_4^-$. Boil the solution until all of the H_2S has been expelled. Add a volume of water equal to that of the solution and add 2 drops of 6 M HCl. Place a piece of recently cleaned aluminum wire about one-eighth inch long in the solution and heat until the wire has completely dissolved. Add 1 drop of 6 M HCl and again heat for a few minutes. If antimony is present, black flakes of the metal will appear. Separate the mixture. Treat the black flakes with 3 drops of 4 M HNO_3 and several drops of 1 M oxalic acid, $H_2C_2O_4$. Add 2 drops of thioacetamide to the solution and place the test tube in a hot water bath. The formation of a red-orange precipitate, Sb_2S_3, confirms the presence of antimony. To the solution obtained from the separation of metallic antimony, add a few drops of 0.2 M $HgCl_2$. The formation of a white, gray, or black precipitate confirms of presence of tin.

The flow sheet for Group II analysis is shown in Table S.3 (p. 412).

Questions and Problems for Group II

1. Write the equation for the hydrogen-ion-catalyzed hydrolysis of thioacetamide. The hydrogen sulfide, as one of the products of the hydrolysis of thioacetamide, is a diprotic acid; illustrate this property of hydrogen sulfide by suitable equations.

2. In terms of ionic equilibria theory, discuss the effect of adding hydrochloric acid on the concentration of the sulfide ion in a solution of hydrogen sulfide. How would ammonia molecules and hydroxyl ions influence the concentration of sulfide ions?

3. Explain in terms of ionic equilibria and solubility product theory, the dissolution of the sulfides of copper, bismuth, cadmium, and lead in nitric acid.

4. Why are the ions of lead and mercury found in both Groups I and II?

5. Explain in terms of ionic equilibria and solubility product theory why Group II is separated from Group III by hydrogen sulfide in the presence of hydronium ions at a concentration of 0.3 M.

6. How does the separation of the Division B sulfides illustrate the amphoteric nature of the sulfides of mercury, arsenic, antimony, and tin?

7. Why is it necessary to remove the nitric acid present before attempting to precipitate lead as the sulfate?

8. Give the color of each of the following: $Cu(H_2O)_4^{2+}$, $Cu(NH_3)_4^{2+}$, $Bi(OH)_3$, CuS, PbS, Bi_2S_3, HgS, As_2S_3, SnS_2, $PbSO_4$, $Cd(NH_3)_4^{2+}$, $Cu(CN)_2^-$, $MgNH_4AsO_4$, Sb, and Bi (finely divided).

9. Explain the dissolution of lead sulfate in ammonium acetate and the reprecipitation of the lead as lead chromate in terms of ionic equilibria and solubility product theory.

10. Why must the sodium stannite that is used in the identification of bismuth be prepared just prior to its use?

11. The Division A sulfides are washed with water containing ammonium nitrate. What is the function of the ammonium nitrate?

12. If a Group II unknown contains copper in an appreciable concentration, this fact should be evident from an inspection of the unknown solution. Why?

13. Select a reagent used in the analysis of Group I or Group II that will in one step separate each of the following pairs: (a) CdS, HgS, (b) Ag^+, Bi^{3+}, (c) Ag^+, Fe^{3+}, (d) As_2S_3, SnS_2, (e) CuS, CdS, (f) Bi^{3+}, Cd^{2+}, (g) $SbCl_4^-$, $SnCl_6^{2-}$.

14. Outline the separation of the following groups of ions, leaving out all unnecessary steps: (a) Ag^+, Hg^{2+}, Co^{2+}, (b) Pb^{2+}, Cu^{2+}, Cd^{2+}, (c) Bi^{3+}, As^{3+}, Sb^{3+}.

15. How will the separation of Groups II and III be affected if the concentration of hydrogen ion in the solution saturated with hydrogen sulfide is 0.1 M? 1 M?

16. Write equations for the test for mercury.

The Analysis of Group III

The solution to be analyzed may be a Group III known or unknown, or it may be the solution from the Group II separation.

Procedure 1. *Precipitation of Group III:* Co^{2+}, Ni^{2+}, Fe^{3+}, Mn^{2+}, Al^{3+}, Cr^{3+}, Zn^{2+}.

(a) If the solution to be analyzed is a known or unknown for Group III only, take 10 drops of the solution, add 1 drop of 6 M HCl, dilute to

TABLE S.3. GROUP II FLOW SHEET

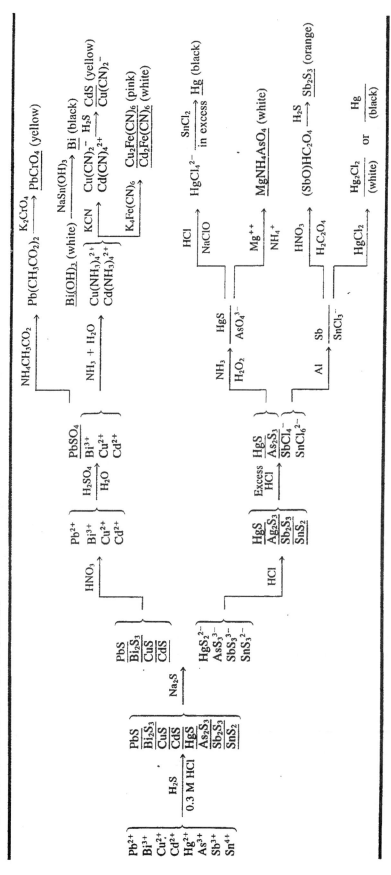

1 mL and add 5 drops of 5% thioacetamide solution. Heat the solution in a hot water bath for at least 5 minutes.

(b) If the solution to be analyzed is that from the Group II separation, add 5 drops of 5% thioacetamide solution and heat the mixture in a hot water bath for at least 5 minutes.

To the solution resulting from either procedure (a) or (b) above, add 5 drops of 15 M aqueous ammonia and stir the precipitate. Heat the mixture for 5 minutes in a hot water bath. Separate the precipitate and wash it with a few drops of water. Reserve the solution for the analysis of Groups IV and V.

Procedure 2. *Precipitate from Procedure 1:* CoS, NiS, FeS, MnS, $Al(OH)_3$, $Cr(OH)_3$, and ZnS. Add 10 drops of 1 M HCl to the precipitate and stir the mixture. Separate the mixture immediately. *Save the centrifugate.* Wash the sulfides that remain (CoS and NiS) with 4 drops of 1 M HCl. Reserve the combined centrifugates for Procedure 4.

Procedure 3. *Residue from Procedure 2:* CoS and NiS. Add 3 drops of 12 M HCl and also 1 drop of 15 M HNO_3 to the residue and heat the mixture in a hot water bath. Separate any sulfur that forms and boil the solution to remove any excess nitric acid or oxides of nitrogen. Add sufficient 4 M aqueous ammonia to the solution to make it *slightly basic* (**important**) to litmus. Dilute solution to 1 mL and divide it into three parts.

(a) *Test for Nickel.* Add 1 drop of dimethylglyoxime to one part of the solution. The formation of a pink or red precipitate confirms the presence of nickel.

(b) *Test for Cobalt.* Acidify a second portion of the solution with 1 M HCl and add several crystals of NH_4SCN to it. Now add an equal volume of acetone and agitate the mixture. The development of a blue color proves the presence of cobalt. If the solution becomes red upon the addition of NH_4SCN, iron(III) ions are present. Add 1 drop of 1 M NaF to the solution. Now, if the solution is bluish-green to green, the presence of cobalt is confirmed.

(c) *Test for Cobalt.* Acidify the third portion of the solution from Procedure 3 with 4 M acetic acid, CH_3CO_2H, and add several large crystals of KNO_2. Warm the mixture. The formation of a yellow precipitate confirms the presence of cobalt.

Procedure 4. *Solution from Procedure 2:* Mn^{2+}, Fe^{2+}, Al^{3+}, Cr^{3+}, and Zn^{2+}. Transfer the solution to a casserole, add 1 mL of 4 M HNO_3, and evaporate the solution to a moist residue. Take up the residue in 1 mL of water and transfer the solution to a test tube. Add 10 drops of 4 M NaOH beyond the amount of this reagent that is required to initiate precipitation. Now add 6 drops of 3% hydrogen peroxide to the mixture and heat it in a hot water bath for 5 minutes. Separate the mixture and wash the residue with 10 drops of water to which has been added 1 drop of 4 M NaOH. Save the combined centrifugates for Procedure 6.

Procedure 5. *Residue from Procedure 4:* $Mn(OH)_3$, MnO_2, and $Fe(OH)_3$. Treat the residue with 1 mL of 4 M HNO_3 and 2 drops of 1 M $NaNO_2$. Stir the mixture and heat it in a hot water bath. Separate any residue that remains. Heat the solution to boiling, then cool it, and divide it into two parts.

(a) *Test for Iron.* Dilute one part of the solution to 1 mL and add 2 or 3 crystals of NH_4SCN. If iron is present, a dark-red color will develop in the solution.

(b) *Test for Manganese.* To the second part of the solution from 5 add a small quantity of solid $NaBiO_3$ and a few drops of 4 M HNO_3. The formation of a pink or purple color that persists confirms the presence of manganese.

Procedure 6. *Solution from Procedure 4:* $Al(OH)_4^-$, CrO_4^{2-}, and $Zn(OH)_4^{2-}$. To the solution add 4 M acetic acid, CH_3CO_2H, until it is acid to litmus and then add 2 or 3 drops of the acid in excess. Now add 4 M aqueous ammonia until the solution is distinctly alkaline to litmus. If a white gelatinous precipitate forms, it is probably aluminum hydroxide. Separate the precipitate and reserve the solution for Procedure 7. Confirm the presence of aluminum by dissolving the precipitate in 4 M acetic acid, CH_3CO_2H, and adding 2 drops of the aluminum reagent and sufficient 1 M $(NH_4)_2CO_3$ to make the solution basic. The formation of a reddish-colored precipitate confirms the presence of aluminum.

Procedure 7. *Solution from Procedure 6:* CrO_4^{2-} and $Zn(NH_3)_4^{2+}$. Divide the solution into two parts.

(a) *Test for Chromium.* To one part of the solution add 1 M acetic acid, CH_3CO_2H, until the solution is acid to litmus. Use red and blue litmus paper for the test. Then add 2 drops of

0.1 M lead(II) acetate, $Pb(CH_3CO_2)_2$. The formation of a yellow precipitate, $PbCrO_4$, confirms the presence of chromium.

(b) *Test for Zinc*. To the other part of the solution add 5 drops of 5% thioacetamide solution and heat the mixture in a hot water bath. The formation of a white precipitate, ZnS, indicates the presence of zinc. Separate, and then dissolve the precipitate in 4 M HCl. Heat the solution to expel the excess hydrogen sulfide and then neutralize it with 4 M aqueous ammonia. Now add 10 drops of 1 M HCl and then 5 drops of 0.1 M $K_4Fe(CN)_6$. The formation of a white precipitate, $K_2ZnFe(CN)_6$, proves the presence of zinc.

The flow sheet for Group III analysis is shown in Table S.4.

Questions and Problems for Group III

1. Why do the sulfides of Co^{2+}, Ni^{2+}, Mn^{2+}, Fe^{2+}, and Zn^{2+} precipitate in an ammoniacal solution of hydrogen sulfide but not in a 0.3 M HCl solution of the gas?
2. What is the function of the ammonium chloride used in the Group III precipitating reagent?
3. Why do aluminum and chromium precipitate as hydroxides rather than as sulfides in Group III?
4. Why does tripositive iron precipitate as iron(II) sulfide rather than iron(III) sulfide in an ammonium sulfide solution?
5. Account for the fact that CoS and NiS fail to precipitate in Group II, yet they dissolve only very slowly in 1 M HCl.
6. A Group III unknown is colorless. What cations are probably absent? Why should one not rely definitely upon such an observation?
7. Cite two cases showing the use of complex ions in the analysis of Group III.
8. Why will $PbCrO_4$ precipitate when Pb^{2+} is added to a solution made up from $K_2Cr_2O_7$?
9. When and why is fluoride added in the test for cobalt using thiocyanate?
10. Write equations showing the amphoteric nature of the hydroxides of zinc, chromium, and aluminum.
11. Show by the proper formulas that manganese can act as either a metal or a nonmetal, depending upon its oxidation state.
12. Select a reagent used in Group III that will separate each of the following pairs: (a) CoS, ZnS, (b) Fe^{3+}, Al^{3+}, (c) $Zn(NH_3)_4^{2+}$, CrO_4^{2-}, (d) Al^{3+}, Zn^{2+}, (e) Mn^{2+}, Mg^{2+}.

13. Outline the separation of the following groups of ions leaving out all unnecessary steps. (a) Hg_2^{2+}, Hg^{2+}, Cu^{2+}, Fe^{2+}, (b) Cd^{2+}, Co^{2+}, Ca^{2+}, (c) Ni^{2+}, Mn^{2+}, Zn^{2+}.
14. Give the color of each of the following: $Fe(OH)_3$, Fe^{3+}, $Fe(OH)_2$, Fe^{2+}, $Fe(SCN)^{2+}$, FeF_6^{3-}.

The Analysis of Group IV

The solution to be analyzed may be a Group IV known or unknown, or it may be the solution from the Group III separation.

Procedure 1. *Precipitation of the Group IV ions:* Ba^{2+}, Sr^{2+}, Ca^{2+}. Evaporate the solution (10 drops of a Group IV known or unknown or the solution from the Group III separation) to dryness and ignite (heat strongly) in a casserole to expel ammonium salts. Dissolve the residue in a mixture of 1 drop of 12 M HCl and 12 drops of water. Make the solution alkaline by adding 4 M aqueous ammonia. Add just 1 drop of aqueous ammonia in excess. Add 2 drops of 1 M $(NH_4)_2CO_3$, or more if necessary to effect complete precipitation, and warm the mixture in a hot water bath. Allow the mixture to cool; separate the precipitate and reserve the solution for the Group V analysis.

Procedure 2. *Precipitate from Procedure 1:* $BaCO_3$, $SrCO_3$, $CaCO_3$. Dissolve the precipitate in a mixture of 2 drops of 4 M acetic acid, CH_3CO_2H, and 4 drops of 1 M ammonium acetate, $NH_4CH_3CO_2$. Add 1 drop of 1 M K_2CrO_4 to the solution. The formation of a yellow precipitate indicates the presence of barium. Separate the mixture and reserve the solution for Procedure 3. Dissolve the precipitate, $BaCrO_4$, in 2 drops of 12 M HCl. Make a flame test on the solution for the barium ion. The barium ion imparts a fleeting greenish-yellow color to the flame. Add 1 drop of 4 M H_2SO_4 to the remainder of the solution. A white precipitate ($BaSO_4$) confirms the presence of barium.

Procedure 3. *Solution from Procedure 2:* Sr^{2+} and Ca^{2+}. Add 4 M aqueous ammonia to the solution until the color changes from orange to yellow. Now add a volume of ethanol equal to the volume of the solution. The formation of a fine yellow precipitate indicates $SrCrO_4$. Separate the mixture and reserve the solution for Procedure 4. Dissolve the precipitate in 2 drops of 12 M HCl and make a flame test on the solution for the strontium ion. A crimson color imparted to the flame is characteristic of the strontium ion.

TABLE S.4. GROUP III FLOW SHEET

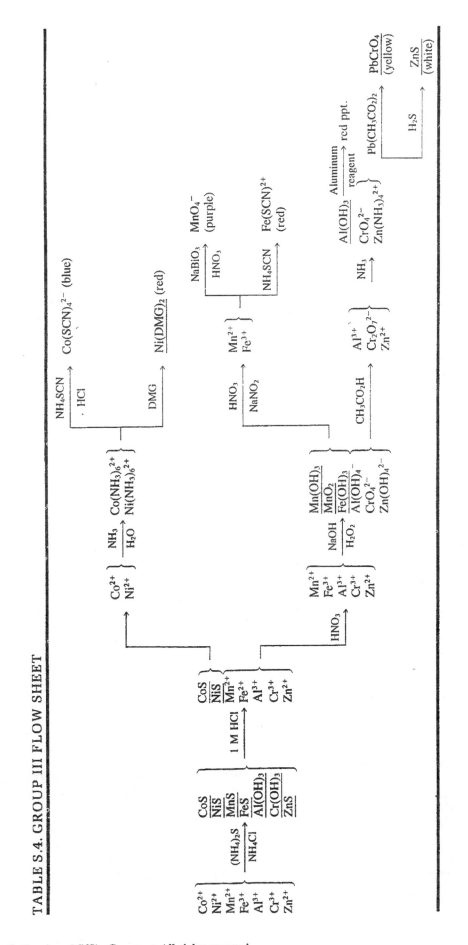

Procedure 4. *Solution from Procedure 3:* Ca^{2+}. Heat the solution to boiling and add 2 drops of 0.4 M $(NH_4)_2C_2O_4$. The formation of a white precipitate confirms the presence of calcium. Dissolve the precipitate in 12 M HCl and make a flame test on the solution. Calcium ions impart a brick-red color to the flame.

Procedure 5. *Original Solution:* Ba^{2+}, Sr^{2+}, Ca^{2+}. Make flame tests on the original solution and compare with tests on known solutions that you make up yourself.

The flow sheet for Group IV analysis is shown in Table S.5.

Questions and Problems for Group IV

1. What reagents compose the Group IV precipitant? What is the function of each of these chemicals in the group precipitation?
2. Write equations representing the various ionic equilibria involved in the Group IV precipitation.
3. Barium chromate is precipitated from a solution of potassium chromate containing acetic acid. Why is the acetic acid present?
4. Why are ammonium salts expelled prior to the precipitation of Group IV?
5. In terms of ionic equilibria and solubility product theory, explain the dissolution of $BaCO_3$ in acetic acid.
6. What characteristic properties cause Ca^{2+}, Sr^{2+}, and Ba^{2+} to be members of the same group of the analytical scheme?
7. What are the characteristic flame color tests for Ca^{2+}, Sr^{2+}, and Ba^{2+}?
8. Outline the separation of the following groups of ions leaving out all unnecessary steps. (a) Ag^+, Cu^{2+}, Zn^{2+}, Ca^{2+}, Na^+, (b) Fe^{3+}, Ba^{2+}, Mg^{2+}, (c) Pb^{2+}, Sn^{4+}, Al^{3+}, Sr^{2+}.
9. Account for the color change from orange to yellow when aqueous ammonia is added to an acetic acid solution of potassium dichromate.
10. Explain why ammonium carbonate rather than sodium carbonate is used in the precipitation of Group IV.

The Analysis of Group V

The solution for analysis may be a Group V known or unknown or it may be the solution for the Group IV separation.

Procedure 1. *Solution from the Group IV separation:* Na^+, K^+, Mg^{2+}, NH_4^+, Ca^{2+}, Ba^{2+}. (Ca^{2+}, Ba^{2+} are present in trace amounts.) Add to this solution 1 drop of 4 M $(NH_4)_2C_2O_4$ and 1 drop of 1 M $(NH_4)_2SO_4$. Separate and discard any precipitate (CaC_2O_4 and $BaSO_4$) that may form and use the solution for Procedure 2.

Procedure 2. *Solution from Procedure 1:* K^+, Na^+, Mg^{2+}, NH_4^+. To 2 drops of the solution from Procedure 1 or Group V known or unknown, add 1 M acetic acid, CH_3CO_2H, until the solution is acid to litmus. Use red and blue litmus paper. Add 1 drop of this acidified solution to 5 drops of sodium reagent. Shake the mixture and set it aside for a while. The formation of a yellow crystalline precipitate indicates the presence of sodium.

Procedure 3. *Use 10 drops of the solution from Procedure 1 or 10 drops of Group V known or unknown:* K^+, Na^+, Mg^{2+}, NH_4^+. Add 4 M aqueous ammonia to the solution until it is alkaline to litmus and then add 1 drop in excess. Now add 2 drops of 1 M Na_2HPO_4 to the solution. The formation of a white crystalline precipitate, often slow in forming, confirms the presence of magnesium. Separate the mixture and save the solution for Procedure 4. Dissolve the precipitate in a mixture of 2 drops of 1 M acetic acid, CH_3CO_2H, and 3 drops of water. Add 1 drop of magnesium reagent and an excess of 4 M NaOH. The formation of a blue precipitate confirms the presence of magnesium. Be *sure* a precipitate forms.

TABLE S.5. GROUP IV FLOW SHEET

Procedure 4. *Solution from Procedure 3:* K^+, Na^+, NH_4^+. Add 4 drops of 15 M HNO_3 to the solution held in a casserole, evaporate it to dryness, and then heat the dry residue for several minutes. After the casserole is cool, add 1 drop of 1 M HCl and 5 drops of 4 M acetic acid, CH_3CO_2H, to the residue and boil the resultant solution. Now add 2 drops of this solution to 5 drops of a saturated solution of $Na_3Co(NO_2)_6$. The formation of a yellow precipitate confirms the presence of potassium.

Procedure 5. *Original known or unknown solution for Group V or solution from Procedure 1.* Make flame tests on this solution and compare with known solutions of the ions in question. Sodium ions impart an intense and lasting yellow color to the flame. A trace of the sodium ion as a contaminant will give a yellow coloration to the flame so do not rely entirely upon the flame test in reporting the presence of sodium ions. Potassium ions give a violet color to a flame. When testing for potassium, observe the flame through a cobalt glass.

Procedure 6. *Original solution of the known or unknown.* Place 10 drops of the solution in a small beaker. Add 4 M NaOH to the solution until it is basic. Moisten pieces of red and blue litmus paper with distilled water and place them on the convex side of a watch glass. Place the watch glass in the usual manner on a beaker containing the solution and warm the solution gently. Avoid spattering of the solution by overheating and do not permit the litmus to come into contact with the solution. If the red litmus turns blue within a short time, and the blue litmus stays blue then the presence of the ammonium ion is confirmed.

The flow sheet for Group V analysis is shown in Table S.6.

Questions and Problems for Group V

1. Why is Group V often referred to as the soluble group?
2. Explain why the test for the ammonium ion must be made on a portion of the original sample during the analysis of either a Group V known or unknown or a general unknown involving all of the groups.
3. Why must the ammonium ion be removed before making the chemical test for the potassium ion?
4. Write the equation for the dissolution of $MgNH_4PO_4$ in an acid. Explain this reaction in terms of solubility product and ionic equilibria theory.
5. Why are $(NH_4)_2SO_4$ and $(NH_4)_2C_2O_4$ added prior to the analysis of Group V?
6. Explain the use of cobalt glass in the flame test for potassium.
7. Outline the chemistry of the chemical test for potassium.
8. Why may red litmus turn blue upon prolonged exposure to the air of the laboratory?

TABLE S.6. GROUP V FLOW SHEET

Analysis for Anions

INTRODUCTION

Thirteen of the more common and important negative ions are included in this scheme of anion analysis. The anions considered are carbonate, sulfide, sulfite, nitrite, sulfate, nitrate, phosphate, metaborate, oxalate, fluoride, chloride, bromide, and iodide. Among the anions that were given consideration in the cation scheme of analysis are arsenite, arsenate, stannite, stannate, permanganate, aluminate, chromate, dichromate, and zincate.

Although several schemes of anion analysis involving the separation of the ions into groups have been developed, the procedures are in general more complicated and unreliable than are those for cation analysis. Instead of using a systematic scheme of analysis on the same solution throughout the analysis, the procedures outlined below involve a series of elimination tests that prove the absence of certain anions. Tests are then made of different samples of the unknown solution for the presence of the anions whose absences were not indicated in the preliminary elimination tests.

The anion knowns and unknowns are in general furnished in the form of mixtures of dry salts because some of the anions react with one another in solution. Thus certain anions may be detected in freshly prepared solutions, whereas, upon standing, these same anions may have undergone decomposition. Certain combinations of anions react almost immediately in alkaline solutions while still other anions cannot exist together in acidic solutions. As an example of anion incompatibility, strong oxidizing anions usually react with strong reducing anions upon acidification.

The general flow sheet for anion elimination is shown in Table 5.7 (p. 420).

Preliminary Elimination Tests for Anions

Remove one of the Anion Elimination Charts from the Appendix of this book, and after you have performed each elimination test, check off in the appropriate spaces those anions that you have found to be absent. When all the elimination tests have been made, it will be apparent which anions may be present. Tests will then be made for the anions that have not been eliminated.

Test 1. *Elimination of anions forming volatile acid anhydrides*, CO_3^{2-}, SO_3^{2-}, S^{2-}, and NO_2^-. Treat 25 mg of the solid unknown contained in a test tube with 2 drops of 1.5 M H_2SO_4. Examine the mixture for the formation of gas bubbles. If no gas is evolved, heat the mixture in the hot water bath.

If no gas is evolved, then CO_3^{2-}, SO_3^{2-}, S^{2-}, and NO_2^- are proved to be absent. If these anions are absent, place a check mark after the formula for each of these anions in the Test 1 column of your anion chart; leave these if a gas is evolved.

If a gas is evolved, **carefully** note its odor and color. Carbon dioxide is colorless and odorless. Sulfur dioxide is colorless and has the odor of burning sulfur. Hydrogen sulfide is colorless and exhibits a characteristic odor. If the nitrite ion is present, nitrogen dioxide, with a reddish-brown color and characteristic odor, is evolved. The solution may be pale blue in color, due to the presence of HNO_2, if the nitrite ion is present.

Preparation of the Solution for Analysis. *Removal of heavy metal ions by* Na_2CO_3. Place

TABLE S.7. ANION ELIMINATION FLOW SHEET

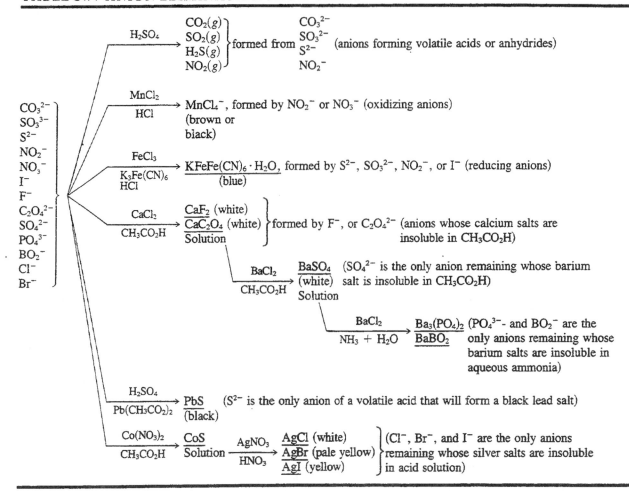

100 mg of the powdered unknown in a test tube and add 2 mL of 1.5 M Na$_2$CO$_3$ solution. Heat the mixture for 10 minutes in the hot water bath. If ammonia is given off, continue the heating until no more gas is evolved. Separate any precipitate that forms. The solution will contain the anions in the form of sodium salts and be referred to as the "prepared solution."

Test 2. *Elimination of oxidizing anions,* NO$_3^-$ and NO$_2^-$. Add 6 drops of a saturated solution of MnCl$_2$ in 12 M HCl to 4 drops of the prepared solution and heat the mixture to boiling. The formation of a dark brown or black coloration, MnCl$_4^-$, indicates the presence of NO$_3^-$, NO$_2^-$, or a mixture of these ions. Record the results of the test on your anion chart.

Test 3. *Elimination of reducing anions,* S^{2-}, SO$_3^{2-}$, I$^-$, and NO$_2^-$. Mix 2 drops of a recently prepared saturated solution of K$_3$Fe(CN)$_6$ with 1 drop of 0.1 M FeCl$_3$ and 2 drops of 6 M HCl. Add 2 drops of the prepared solution to this mixture

and let stand for a few minutes. The development of a blue coloration or the formation of a blue precipitate indicates the presence of a reducing agent (S^{2-}, SO$_3^{2-}$, I$^-$, NO$_2^-$). Run a blank test with the same reagents but leave out the prepared solution and compare the colors and intensities of colors of the unknown and blank test solutions.

Test 4. *Elimination of fluoride and oxalate,* F$^-$ and C$_2$O$_4^{2-}$. To 20 drops of the prepared solution add 4 M acetic acid (CH$_3$CO$_2$H) (count the drops) until the solution is just acid to red and blue litmus paper. Now add an equal number of drops of 4 M acetic acid (CH$_3$CO$_2$H) in excess. Tap the test tube repeatedly until the excess of CO$_2$ from the Na$_2$CO$_3$ present is expelled. Add 8 drops of 0.1 M CaCl$_2$ and shake the contents of the tube. If a precipitate forms after a few minutes, it may be CaF$_2$, CaC$_2$O$_4$, or a mixture of these compounds. If no precipitate forms, F$^-$ and C$_2$O$_4^{2-}$ are absent. Separate the precipitate and use the solution for Test 5. Save the precipitate for the confirmatory test for C$_2$O$_4^{2-}$.

Test 5. *Elimination of sulfate:* SO_4^{2-}. Test for completeness of precipitation in the solution from Test 4 by adding 1 drop of 0.1 M $CaCl_2$. Now add 8 drops of 0.1 M $BaCl_2$. Shake the mixture and let it stand for 5 minutes. If SO_4^{2-} is present, a precipitate of $BaSO_4$ will form. Separate and reserve the solution for Test 6. If SO_3^{2-} was present in the original solution, you may obtain a test for SO_4^{2-} due to the ease with which the former is oxidized to the latter by oxygen of the air.

Test 6. *Elimination of phosphate and metaborate,* PO_4^{3-} *and* BO_2^-. Transfer the solution from Test 5 to the casserole, add 6 drops of 12 M HCl, and heat to expel SO_2 and CO_2. Make the solution alkaline with aqueous ammonia and then add 5 drops in excess. A white precipitate may be $Ba_3(PO_4)_2$, $Ba(BO_2)_2$, or a mixture of these. Because barium metaborate is prone to form supersaturated solutions, it is necessary to carry out the confirmatory test for BO_2^- even though no precipitate forms in Test 6.

Test 7. *Elimination of sulfide,* S^{2-}. To 4 drops of the prepared solution contained in a small beaker add sufficient 1.5 M H_2SO_4 to make the solution acidic to litmus paper. Cover the beaker with a watch glass to the under side of which is attached a small piece of moist lead acetate paper. The appearance of a black stain on the paper after a few minutes indicates the presence of the S^{2-} ion.

Test 8. *Elimination of chloride, bromide, and iodide,* Cl^-, Br^-, *and* I^-. If the sulfide ion is present, treat 10 drops of the prepared solution with 4 M CH_3CO_2H until acid to litmus paper. Now add dropwise with shaking 10 drops of 1 M $Co(NO_3)_2$. Heat in the hot water bath for a few minutes and separate the precipitate of CoS. The solution may contain Cl^-, Br^-, and I^-. To the solution, or if sulfide is absent, to 10 drops of the prepared solution acidified with 1 M HNO_3, add 5 drops of 1 M HNO_3. Now add to this solution 0.1 M $AgNO_3$ until precipitation is complete. Separate and wash the precipitate with 10 drops of water. Discard the solution. Mix 10 drops of 0.1 M $AgNO_3$, 6 drops of aqueous ammonia, and 4 mL of water. Add this mixture to the precipitate and stir thoroughly. If the precipitate dissolves completely, Br^- and I^- are absent, and Cl^- is indicated. Acidify the solution with 4 M HNO_3. A white precipitate confirms chloride. If the precipitate did not dissolve completely in the aqueous ammonia and silver nitrate mixture, it may have partially dissolved. Separate and acidify the solution

with 4 M HNO_3. If no precipitate forms, Cl^- is absent. The formation of a heavy white precipitate confirms the presence of Cl^-. A slight precipitate or turbidity, yellow in color, may be silver bromide.

Confirmatory Tests for the Anions

Procedure 1. *Identification of* CO_3^{2-}. To 25 mg of the powdered solid unknown add 10 drops of 3% hydrogen peroxide and heat the mixture in the hot water bath. Place a drop of $Ba(OH)_2$ solution in a loop of platinum wire. Add 3 drops of 4 M H_2SO_4 to the unknown mixture and immediately hold the drop of $Ba(OH)_2$ over the reaction mixture. The formation of a definite turbidity in the $Ba(OH)_2$ solution indicates the presence of CO_3^{2-}

Procedure 2. *Identification of* SO_4^{2-}. Acidify 5 drops of the prepared solution with 6 M HCl and add 2 drops in excess. Heat in the hot water bath to expel the excess of CO_2. Add 0.1 M $BaCl_2$ until precipitation is complete; then allow the mixture to stand. A white, finely divided precipitate of $BaSO_4$ confirms the presence of SO_4^{2-}. Separate and save the solution for the SO_3^{2-} test.

Procedure 3. *Identification of* SO_3^{2-}. Add 5 drops of bromine water and 2 drops of 0.1 M $BaCl_2$ to the solution from Procedure 2. Heat in the hot water bath and allow to stand for 5 minutes. A white precipitate, $BaSO_4$, shows the presence of sulfite in the original solution.

Procedure 4. *Identification of* S^{2-}. The sulfide ion is confirmed in elimination Test 7.

Procedure 5. *Identification of* NO_2^-. To 5 drops of prepared solution add 4 M H_2SO_4 dropwise until the solution is acidic. Now add 5 drops of freshly prepared 0.1 M Fe SO_4 solution. If NO_2^- is present the solution will assume a dark brown coloration (NO_2 gas is evolved).

Procedure 6. *Identification of* NO_3^- *in the absence of* NO_2^-, Br^-, *and* I^-. To 5 drops of prepared solution in a test tube add 4 M H_2SO_4 dropwise until the solution is acidic. Now add 5 drops of freshly prepared 0.1 M $FeSO_4$ solution. Add 5 drops of 18 M H_2SO_4 (**Caution!**) by holding the test tube in an inclined position so that the sulfuric acid can run down the side of the test tube and form a separate layer on the bottom of the tube. If the NO_3^- is present, a brown ring will form at the junction of the two liquids within a few minutes.

Procedure 7. *Identification of NO_3^- in the presence of NO_2^-. Removal of NO_2^-.* To 6 drops of the prepared solution add 4 M H_2SO_4 until acidification occurs (test with litmus) and then add 4 drops of 1 M $(NH_4)_2SO_4$ solution. Place the mixture in a casserole and slowly evaporate the solution to moist residue (not dryness). Add 4 drops of water and evaporate to a moist residue a second time. Dissolve the residue in 10 drops of water and transfer the mixture to a small test tube. Repeat the brown ring test for NO_3^- as described in Procedure 6 on the mixture resulting from the removal of NO_2^-.

Procedure 8. *Identification of NO_3^- in the presence of Br^- and I^-. Removal of Br^- and I^-.* To a test tube containing 6 drops of the prepared solution, add 10 drops of water. Acidify the solution (test with litmus) with 4 M CH_3CO_2H and then add 80 mg of powdered Ag_2SO_4 (NO_3^- free). Stir and grind the mixture in the test tube for 2–3 minutes. Separate and transfer the solution to a test tube. Repeat the brown ring test for NO_3^- as described in Procedure 6 on the solution obtained from the removal of Br^- and I^-

Procedure 9. *Identification of $C_2O_4^{2-}$.* Wash the precipitate obtained in preliminary elimination Test 4 twice with 10 drops of water each time and discard the washings. To the residue add 10 drops of water and 10 drops of 4 M H_2SO_4, and shake the mixture. Add 0.002 M $KMnO_4$ dropwise until 1 drop imparts a permanent pink color to the solution. If $C_2O_4^{2-}$ is present, several drops of $KMnO_4$ should be required to give a permanent pink coloration to the solution. Run a blank by repeating the test but leaving out the precipitate from Test 4 of the elimination tests.

Procedure 10. *Identification of F^-.* Place 25 mg of powdered sample in a test tube and add an equal volume of powdered silica. Select a cork to fit the test tube. Cut a notch on the side and a hole in the center of the cork. Insert a glass rod, to one end of which has been sealed a platinum wire looped at the end. Add 2 drops of 18 M H_2SO_4 **(Caution!)** to the mixture in the tube. Now place the platinum wire loop with a drop of water hanging to it in the test tube directly above the reaction mixture. Warm the test tube in the hot water bath and then set it aside to cool. The formation of a white precipitate or cloudiness in the drop of water confirms the presence of the F^- ion.

Procedure 11. *Identification of PO_4^{3-}.* Dilute 4 drops of prepared solution with 10 drops of water. Make the solution just acidic with 4 M HNO_3 and add 10 drops of Magnesia mixture. A white precipitate, often slow in forming, confirms the presence of the PO_4^{3-} ion.

Procedure 12. *Identification of BO_2^-.* Evaporate 6 drops of the prepared solution to a small volume in the casserole. Add 1 mL of methanol and 5 drops of 18 M H_2SO_4. **(Caution!)** Transfer the mixture to a test tube and place the tube in the hot water bath. When the alcohol begins to boil, ignite the vapors. A green tinge to the flame confirms the presence of the borate ion.

Procedure 13. *Identification of I^-.* Dilute 6 drops of the prepared solution with 12 drops of water. Add 4 M HNO_3 until the solution is just acid to litmus paper and then add two drops of the acid in excess. Treat the solution with 1 mL of 0.1 M $Fe(NO_3)_3$ and 10 drops of dichloromethane, CH_2Cl_2, and shake the test tube. The development of a violet color in the CH_2Cl_2 layer proves the presence of I^-. Remove the CH_2Cl_2 layer by means of a capillary syringe and discard it. Add 10 more drops of CH_2Cl_2, shake, and remove the CH_2Cl_2 layer. Repeat the extraction until the CH_2Cl_2 layer remains colorless and use the aqueous solution for the identification of Br^-.

Procedure 14. *Identification of Br^-.* Add to the solution from Procedure 13 two drops of 4 M HNO_3, then 0.1 M $KMnO_4$ solution drop by drop until the solution remains pink. Extract the solution with CH_2Cl_2. A yellow or orange coloration in the CH_2Cl_2 layer indicates the presence of Br^-

Questions and Problems for Anions

1. In terms of solubility product and ionic equilibria theory, explain the following: (a) $Ca_3(PO_4)_2$ is soluble in HCl while $BaSO_4$ is not; (b) $CaCO_3$ is soluble in CH_3CO_2H while CaC_2O_4 is not.

2. Explain why the nitrite ion will respond to both the test for oxidizing agents and that for reducing agents.

3. If an anion unknown contains the sulfite ion, one may obtain a test for the sulfate ion even though sulfate was not used in making up the unknown. Explain.

4. What is meant by "incompatibility of anions"? Give an example.

5. What principle is involved in the tests for the halide ions in the presence of one another?

6. Explain how the nitrite ion interferes with the test for nitrate. How is the nitrite ion removed prior to the test for nitrate?

7. Make a list of anions that are reducing agents and list their oxidation products.

8. Make a list of anions that are oxidizing agents and list their reduction products.

9. If a solution is strongly acidic, what anions cannot be present in appreciable concentrations?

10. An acidic solution contains silver ions. What anions are probably absent?

11. A neutral solution contains calcium ions. What anions are probably absent?

12. A solution is slightly acid and contains sulfide ions. What cations are probably absent?

13. Why is it necessary to extract iodine in the confirmatory test for the iodide ion?

14. Outline a simple test to distinguish between each of the following pairs: (a) CO_3^{2-} and SO_4^{2-}, (b) Cl^- and I^-, (c) F^- and BO_2^-, (d) PO_4^{3-} and Br^-, (e) S^2 and SO_3^{2-}, (f) $C_2O_4^{2-}$ and Cl^-.

15. A solution known to contain either Na_2CO_3 or Na_2SO_4 turns red litmus paper blue. Which salt is present?

Analysis of Solid Substances

INTRODUCTION

The analytical procedures that have been outlined for cations and anions on the preceding pages are usually applicable to solutions of the substances being analyzed. However, many substances are not readily soluble in water or even in acids. Thus it becomes necessary to use special procedures for effecting the solution of certain solid substances before proceeding with the analyses.

The Dissolution of Nonmetallic Solids

Procedure 1. *Sample is soluble in water.* Treat 100 mg of the solid with 1 mL of water. If no dissolution is apparent, heat the mixture in the hot water bath for a few minutes. If none of the sample appears to dissolve, evaporate a few drops of the supernatant liquid on a watch glass to determine whether partial dissolution has occurred. If a solid residue remains on the watch glass, heat the sample with fresh portions of water until complete dissolution has been effected. Concentrate the combined extracts by evaporation and proceed with the analysis for cations, anions, or both.

Procedure 2. *Sample is not soluble in water.* Warm a small sample of the solid substance with 6 M HCl or, if this has no effect, with 12 M HCl. (**Caution!**) If solution does not occur, or if it is incomplete, try 6 M HNO_3, 15 M HNO_3 (**Caution!**), and aqua regia (1 part 15 M HNO_3 to 3 parts 12 M HCl (**Caution! Use in hood),** in succession on small samples of the solid until a suitable solvent is found. If dissolution has been effected by any of these reagents, dissolve 100 mg of the sample in the solvent selected. Evaporate the solution to incipient dryness and then take up the residue in 2 mL of water. Use this solution for the analytical procedures.

Procedure 3. *Treatment of the residue with Na_2CO_3 solution.* If HCl, HNO_3, or aqua regia does not completely dissolve the solid, the residue remaining after the acid treatment may be treated with a Na_2CO_3 solution. Treat the residue with 2 mL of 1.5 M Na_2CO_3 in a casserole and boil the mixture for 10 minutes, replacing the water that is lost during evaporation. This procedure will convert many insoluble salts such as $BaSO_4$, $CaSO_4$, $PbSO_4$, and many oxides into insoluble carbonates. Separate the mixture and discard the solution unless it is to be used for anion analysis. Wash the precipitate with water and then dissolve it in a few drops of 6 M HNO_3. Dilute the solution with 1 mL of water and add it to the original acid solution that is to be evaporated. If solution of the residue was not complete, repeat the Na_2CO_3 treatment. The halides of silver, silicate salts, some oxides, and calcined salts are not dissolved by this procedure.

Procedure 4. *Reduction of silver halides with zinc.* Suspend the residue containing the insoluble silver halides in 5 drops of water, then add 1 mL of 1 M H_2SO_4 and a few granules of zinc metal. Warm, and stir the mixture for several minutes. **Avoid direct flames.** Add more zinc if the evolution of hydrogen ceases. Separate the residue of precipitated silver. Wash the residue with water, dissolve it in 6 M HNO_3, and test the solution for the silver ion. Conduct tests for the halide ions on the solution from the silver separation.

Procedure 5. *Fusion with sodium carbonate.* Silicates and certain oxides and calcined salts may not be taken into solution by treatment with acid or Na_2CO_3 solution, or by reduction with zinc. Fusion with Na_2CO_3 is effective with many of these substances. Transfer the residue remaining after the zinc reduction of silver halides to a small nickel crucible. Add to the residue 100 mg of anhydrous Na_2CO_3, about half as much K_2CO_3, and a few milligrams of $NaNO_3$. Place the crucible in a small clay triangle and then heat it in the hot flame of a Meker burner (see Fig. S.7) until the mixture has fused. Cool the crucible, add 1 mL of water and warm until the solid mass has disintegrated. Separate the mixture and reserve the solution. Treat the residue with 8 drops of 6 M HNO_3 and warm in the hot water bath for a few minutes. Separate and add this nitric acid solution to the original solution. Analyze the solution for its cations.

Analysis of Metals and Alloys

All of the common metals except Al, Cr, Mn, Fe, Sb, and Sn may be taken into solution with dilute nitric acid. Al, Cr, Mn, and Fe are attacked superficially by nitric acid with the formation of an oxide film that makes solution of these metals too slow to be practical. Metastannic acid, H_2SnO_3, and the insoluble oxides of antimony, Sb_4O_6 and Sb_2O_5, are formed when nitric acid reacts directly with these metals. Although these compounds are insoluble in nitric acid, they may be readily converted to the corresponding sulfides, which are soluble in nitric acid. Hydrochloric acid is in general not suitable as a solvent for alloys of unknown composition since the volatile hydrides of sulfur, arsenic, phosphorus, and antimony may be formed and consequently these elements could escape detection. On the other hand, nitric acid oxidizes these elements of sulfate, arsenate, phosphate, and antimony oxides, which are nonvolatile. Aqua regia, a mixture of nitric and hydrochloric acids,

will generally dissolve alloys that resist the action of nitric acid alone. **Reminder: Goggles and apron must be worn for Procedures 1–5.**

Procedure 1. *Preparation of the sample for dissolution.* Convert the metal sample into a finely divided state in order to produce a large surface for interaction with the solvent. This may be accomplished by means of a steel file, a mortar and pestle for brittle metals, a hammer for malleable metals, or a knife for soft metals.

Procedure 2. *Selection of a suitable solvent.* Treat a small sample of the metal with 6 M HNO_3 in a test tube and warm the mixture if necessary. If the sample reacts completely with or without the formation of a white precipitate, proceed to Procedure 3. If the sample does not dissolve readily, try aqua regia (**Caution! Use in hood**) (1 part of 15 M HNO_3 and 3 parts of 12 M HCl). If aqua regia fails to attack the sample, it may be treated with a mixture of concentrated HCl and Br_2, or fused with solid NaOH in a silver crucible.

Procedure 3. *Dissolution by HNO_3.* Place 20 mg of the sample in a test tube, add 1 mL of 6 M HNO_3, and heat the mixture in the hot water bath. If a white residue forms, stir the mixture to remove the coating from the surface of the undissolved metallic particles. Add more HNO_3 if required to complete reaction. After the metal has dissolved, transfer the solution or mixture to a casserole and evaporate it to incipient dryness. Add 5 drops of 15 M HNO_3 and again evaporate to incipient dryness. Now add 5 drops of 6 M HNO_3 and 1 mL of water and transfer to a test tube. If a clear solution is obtained, analyze the solution according to the procedures for the cations. If there is a residue (H_2SnO_3, Sb_2O_5, or a mixture of these), separate and analyze the solution according to the procedures for the cations, omitting the tests for tin and antimony. Add 10 drops of NaOH and 5 drops of 5% thioacetamide solution (a source of H_2S) to the residue. Heat the mixture in the water bath for 5 minutes. Separate any residue and analyze the solution for tin and antimony.

Procedure 4. *Dissolution by aqua regia.* (**Caution! Use in hood.**) Place 20 mg of the sample in a test tube, and add 10 drops of 15 M HNO_3 and 30 drops of 12 M HCl. Heat the mixture in the hot water bath until the reaction is complete. Transfer the reaction mixture to a casserole and evaporate it to a small volume (not to dryness). When cool, add 2 mL of water and 4 drops of 6 M HCl. Sepa-

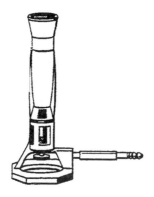

Figure S.7. Meker burner.

rate any precipitate that forms. Analyze the centrifugate, or clear solution if no precipitate forms, by the procedures for Groups II, III, IV, and V. The precipitate may consist of AgCl, PbCl$_2$, and SiO$_2$. Analyze the precipitate for Group I.

Procedure 5. *Dissolution by Br$_2$ and HCl.* If neither HNO$_3$ nor aqua regia will dissolve the metallic sample and a preliminary test has shown that Br$_2$ and HCl will do so, place 20 mg of the sample in a casserole. Add 1 mL of 12 M HCl and then a few drops of liquid bromine to the sample. **(Caution! Use in hood.)** Heat the mixture gently until reaction is complete and then evaporate it to dryness. Add 5 drops of 6 M HNO$_3$ to the residue, dilute with 2 mL of water, and then proceed with the cation analysis.

Appendix

Units and Conversion Factors

Base units of international system of units (SI)		
Physical property	Name of unit	Symbol
Length	Meter	m
Mass	Kilogram	kg
Time	Second	s
Electric current	Ampere	A
Thermodynamic temperature	Kelvin	K
Luminous intensity	Candela	cd
Quantity of substance	Mole	mol

Units of length

Meter (m) = 39.37 inches (in) = 1.094 yards
Centimeter (cm) = 0.01 m
Millimeter (mm) = 0.001 m
Kilometer (km) = 1000 m
Angstrom unit (Å) = 10^{-8} cm = 10^{-10} m

Yard = 0.9144 m
Inch = 2.54 cm

Mile (U.S.) = 1.609 km

Units of volume

Liter (L) = 0.001 m^3 = 1000 cm^3
Milliliter (mL) = 0.001 liter = 1 cm^3

Liquid quart (U.S.) = 0.9463 liter
Cubic foot (U.S.) = 28.316 liters

Units of weight

Gram (g) = 0.001 kg

Milligram (mg) = 0.001 g
Kilogram (kg) = 1000 g
Ton (metric) = 1000 kg = 2204.62 lb

Ounce (oz) (avoirdupois) = 28.35 g
Pound (lb) (avoirdupois) = 0.45359237 kg
Ton (short) = 2000 lb = 907.185 kg
Ton (long) = 2240 lb = 1.016 metric tons

Units of energy

4.184 joule (J) = 1 thermochemical calorie (cal) = 4.184 × 10^7 erg
Erg = 10^{-7} J
Electron volt (eV) = 1.6021 × 10^{-12} erg = 23.061 kcal/mol
Liter atmosphere = 24.217 cal = 101.32 J

Unit of force

Newton (N) = 1 $kg \cdot m \cdot s^{-2}$ (force which when applied for 1 sec will give to a 1-kilogram mass a speed of 1 meter per second)

Units of pressure

Torr = 1 mm Hg
Atmosphere (atm) = 760 mm Hg = 760 torr = 101,325 $N \cdot m^{-2}$ = 101,325 Pa
Pascal = $kg \cdot m^{-1} \cdot s^{-2}$ = $N \cdot m^{-2}$

Dimensional Analysis

Many problems can be solved using an approach known as *dimensional analysis* (sometimes referred to as the factor-label or unit-factor method). This technique is based on the dimensions (units) of the quantities involved in a calculation, such as inches, centimeters, grams, or moles. This method is used to convert from one set of units to another.

Step 1 in using this technique is to employ a *conversion factor*, which is essentially a ratio relating two different units; that is, expressing the desired quantity in two different ways. The following conversion factor is an example relating inches and feet:

$$12 \text{ inches} = 1 \text{ foot}$$

Every conversion factor yields two quotients, each having the value 1, since the quantity on the top of each fraction is equal to that on the bottom of the fraction. Thus the above conversion factor relating inches and feet may be written as follows:

$$\frac{12 \text{ inches}}{1 \text{ foot}} = 1 \quad \text{or} \quad \frac{1 \text{ foot}}{12 \text{ inches}} = 1$$
$$\text{Quotient (1)} \qquad\qquad \text{Quotient (2)}$$

It is therefore evident that multiplying another quantity by a conversion factor in its proper quotient form will not change its size, only its units.

Step 2 in using this technique is to take the quantity to be converted *together with its unit* and multiply it by the quotient that will yield the desired units in the answer. For example, to convert 150 feet to inches, take the quantity 150 feet and multiply it by Quotient (1):

$$150 \text{ feet} \times \frac{12 \text{ inches}}{1 \text{ foot}}$$

$$= \frac{(150)(12)}{1} \text{inches} = 1800 \text{ inches}$$

This procedure results in the answer with the desired units; namely, *inches*.

On the other hand, to convert 150 inches to feet, employ Quotient (2) in order to obtain the answer in the desired units, in this case, *feet*:

$$150 \text{ inches} \times \frac{1 \text{ foot}}{12 \text{ inches}}$$

$$= \frac{(150)(1)}{12} \text{feet} = 12.5 \text{ feet}$$

Dimensional analysis is also useful in other types of conversions, such as changing grams to moles, or moles to grams.

For example, we want to solve the problem "convert 15 *g of Na* to *moles of Na.*"

Step 1: Use a conversion factor relating *grams of Na* and *moles of Na,* which is

$$23 \text{ g Na} = 1 \text{ mol Na}$$
$$\text{(1 GAW)}$$

Step 2: Take the quantity to be converted together with its unit and multiply it by the proper quotient obtainable from the selected conversion factor (given above):

$$15 \text{ g Na} \times \frac{1 \text{ mol Na}}{23 \text{ g Na}}$$

$$= \frac{15}{23} \text{mol Na} = 0.65 \text{ } mol \text{ Na}$$

A similar process would be used to convert 0.65 *mole of Na* to *grams of Na:*

Step 1: Use the same applicable conversion factor given in the above example, which relates grams of Na and moles of Na.

Step 2: Take the quantity to be converted together with its unit and multiply it by the proper quotient:

$$0.65 \text{ mol Na} \times \frac{23 \text{ g Na}}{1 \text{ mol Na}}$$

$$= \frac{(0.65)(23)}{1} \text{g Na} = 15 \text{ g Na}$$

Accuracy and Precision

It is important for scientists to know how well they have made their measurements and how dependable the measurements are. If the correct result is known, they need only to compare their result with the known one. If they determine the atomic mass of silver as 107.604, for instance, and the accepted value is 107.870, there is an error of 0.266 atomic mass unit. The relative error is 2.5 parts per 1000, or 0.25%, and is an estimate termed *accuracy*. In practically all situations, however, the correct value is not known, and scientists must depend on the agreement among a series of results made under the same conditions to tell how correct the value is. Suppose five determinations of molecular mass are made on the same substance by the same method and the values 61.2, 60.7, 62.2, 60.1, and 62.0 are obtained. The average is 61.2. To evaluate the result, however, the agreement among the measurements should also be known. Obviously, 82.3 and 40.1 would also average 61.2, and yet one could place little reliance on this average.

A simple way of arriving at the agreement within a set of measurements is to compute the average deviation of the results from the average. In the case above the deviations are 0.0, 0.5, 1.0, 1.1, and 0.8, and the average deviation is 0.7 from 61.2; this is 11 parts per thousand, or 1.1%, and this is the *precision*. Clearly, the smaller the spread among the measurements, the more certain the scientist of their reliability. A common way of expressing both the average value and its precision is to indicate the deviation as a $\pm$ value, here as 61.2 ± 0.7.

The complete mathematical treatment of errors is extensive, but the following simplified discussion should enable you to distinguish between the accuracy and precision of your measurements. Assume that a student runs three density determinations on an object of a known density of 7.53 g/mL and obtains the following results: 7.69, 7.59, and 7.62 g/mL. The mean (average) is found to be 7.63 g/mL, simply by adding the three values and dividing by 3. The accuracy of the student's determination (a measure of systematic errors); that is, agreement of the mean (7.63 g/mL) with the known value (7.53 g/mL), is 98.7%, since the mean value differs from the known by 0.10 g/mL in 7.53 g/mL.

The precision of the student's determination (a measure of random errors) may be expressed in terms of the *standard deviation s* by a statistical method mathematically related to the bell-shaped (Gaussian) probability curve, determined by using the following equation:

$$s = \sqrt{\frac{\sum_i (x_i - \bar{x})^2}{n - 1}}.$$

where x_i is the ith measured value, $\bar{x}$ is the mean value, and n is the number of individual values. Applying the above equation to our example of a student's determination of density, the standard deviation s may be calculated by substituting the values into the equation

$$s = \sqrt{\frac{(7.69 - 7.63)^2 + (7.59 - 7.63)^2 + (7.62 - 7.63)^2}{3 - 1}}$$

$$= \sqrt{\frac{(0.06)^2 + (-0.04)^2 + (-0.01)^2}{2}}$$

$$= \sqrt{\frac{(0.0036) + (0.0016) + (0.0001)}{2}}$$

$$= \sqrt{\frac{(0.0053)}{2}} = \sqrt{(0.0026)} = 0.05 \text{ g/mL}$$

Thus the precision of the student's determination, expressed in terms of the standard deviation, is 0.05 g/mL in 7.63 g/mL; this reperesents 99.3% agreement among the separate values.

These and other methods of evaluating data are available through statistical analysis. Books on this subject should be consulted for details if desired. It is important for you at this point to be able to distinguish, in general, between the meaning of the terms accuracy and precision in the evaluation of your data.

Significant Figures

Associated with every experimental measurement is a degree of uncertainty whose magnitude depends upon the nature of the measuring device and the skill with which it is used. It is meaningless to use more figures than the data justify. It is understood that there is an uncertainty of at least one unit in the last digit of a measurement. For example, a volume measurement of 7 mL using a large graduated cylinder may be reported as 7 ± 1 mL; 7.0 mL using a small graduated cylinder as 7.0 ± 0.1 mL; 7.00 mL using a buret as 7.00 ± 0.01 mL. The method of specifying the degree of confidence in a measurement is often described in terms of *significant figures*. Thus in 7.00 mL there are three significant figures; in 7.0 mL there are two significant figures; in 7 mL there is one significant figure.

The digit zero is always significant unless it is merely marking the place of the decimal, in which case it may or may not be significant. In the case of a number less than one, the zeros preceding the first nonzero digit are never significant, but any zeros to the right of this digit would be significant. For numbers of 10 or more the significance of the zero is ambiguous. This ambiguity is completely removed by using standard exponential notation. Thus if 1000 is reported as 1.00×10^3 it is clear that there are three, rather than four, significant figures—the ambiguity is removed.

● EXAMPLES: The following numbers all contain four significant figures:

1030 (1.030×10^3)
1.030×10^4
0.01030
0.001030

When experimental quantities are multiplied or divided, the number of significant figures in the result cannot exceed that in the least precise measurement.

● EXAMPLES:

$21.02 \times 0.0021 = 0.044$ (not 0.044142)

$6.52 \times 14 \times 3.50 = 3.2 \times 10^2$ (not 319.4800)

When experimental quantities are added or subtracted, the final result should contain the same number of decimal places as the component with the fewest decimal places.

● EXAMPLES:

$$\begin{array}{r} 11.21 \\ 0.2 \\ 256 \\ \hline 267 \end{array} \quad \text{(not 267.41)}$$

$$\begin{array}{r} 25.6 \\ 3.87 \\ 7.739 \\ \hline 37.2 \end{array} \quad \text{(not 37.209)}$$

Rounding Off Numbers

When rounding off numbers, observe the following rules:

1. If the first digit to be dropped is less than 5, leave the last retained digit unchanged.
2. If the first digit to be dropped is greater than 5 or 5 followed by any nonzero digits, increase the last retained digit by 1.
3. If the first digit to be dropped is exactly 5, round off the result to the nearest even number.

● EXAMPLES: When rounded off to four figures, the following numbers all become 27.54:

27.5382   27.5450
27.53501   27.5350
27.5449

Standard Exponential (Scientific) Notation

Standard exponential (scientific) notation is a convenient way of expressing very large and very small numbers. The method consists of expressing the number as a product of two numbers: the first number is the digit term, and is a number not less than 1 and not greater than 10; the second number is the exponential term to the base 10.

● EXAMPLES:

$$1000 = 1 \times 10^3$$
$$100 = 1 \times 10^2$$
$$10 = 1 \times 10^1$$
$$1 = 1 \times 10^0$$
$$0.1 = 1 \times 10^{-1}$$
$$0.01 = 1 \times 10^{-2}$$
$$0.001 = 1 \times 10^{-3}$$
$$0.0001 = 1 \times 10^{-4}$$
$$2345 = 2.345 \times 10^3$$
$$0.123 = 1.23 \times 10^{-1}$$

For a detailed, programmed treatment of algebraic operations involving exponential numbers (and logarithms), see *Solutions for General Chemistry and General Chemistry with Qualitative Analysis, Tenth Editions,* by Meiser; also *Complete Solutions* by Meiser.

Oxidation States and Ionic Charges of Some Common Elements and Ions

+1   Ag, Br, Cl, Cu, H, I, K, Na, NH_4^+
+2   Ba, Ca, Cd, Co, Cr, Cu, Fe, Hg, Hg_2^{2+}, Mg, Mn, Ni, Pb, Sn, Sr, Zn
+3   Al, As, Bi, Cr, Fe, Mn, N, P, Sb
+4   C, Mn, Pb, S, Si, Sn
+5   As, Bi, Br, Cl, I, N, P, Sb
+6   Cr, Mn, S
+7   Cl, I, Mn
−1   Br, $C_2H_3O_2^-$, Cl, ClO^-, ClO_2^-, ClO_3^-, ClO_4^-, F, H, HCO_3^-, HSO_4^-, I, MnO_4^-, NO_2^-, NO_3^-, OH^-, SCN^-
−2 CO_3^{2-}, CrO_4^{2-}, $Cr_2O_7^{2-}$, O, O_2^{2-}, S, SiO_3^{2-}, SO_3^{2-}, SO_4^{2-}, $S_2O_3^{2-}$
−3 AsO_3^{3-}, AsO_4^{3-}, N, P, PO_4^{3-}
−4   C, Si

Density of Water at Various Temperatures

Temp. °C	Density, g/cm^3	Temp. °C	Density, g/cm^3
20.0	0.99820	24.0	0.99730
20.2	0.99816	24.2	0.99725
20.4	0.99812	24.4	0.99720
20.6	0.99808	24.6	0.99715
20.8	0.99804	24.8	0.99710
21.0	0.99799	25.0	0.99704
21.2	0.99795	25.2	0.99699
21.4	0.99790	25.4	0.99694
21.6	0.99786	25.6	0.99689
21.8	0.99782	25.8	0.99684
22.0	0.99777	26.0	0.99678
22.2	0.99772	26.2	0.99673
22.4	0.99768	26.4	0.99668
22.6	0.99763	26.6	0.99662
22.8	0.99759	26.8	0.99657
23.0	0.99754	27.0	0.99651
23.2	0.99749	27.2	0.99646
23.4	0.99744	27.4	0.99640
23.6	0.99739	27.6	0.99635
23.8	0.99735	27.8	0.99629

Solubility Rules for Salts in Water

The following generalizations apply to simple compounds of the more common metals. Exceptions occur for compounds of other metals.

1. All nitrates and acetates are soluble.
2. All chlorates are soluble; potassium chlorate is slightly soluble.
3. All chlorides are soluble except those of mercury(I), silver, lead, and copper(I); lead chloride is soluble in hot water.
4. All sulfates, except those of strontium, barium, and lead are soluble; calcium sulfate and silver sulfate are slightly soluble.
5. Carbonates, phosphates, borates, arsenates, and arsenites, except those of ammonium and the alkali metals, are insoluble.
6. Only the sulfides of ammonium, the alkali metals, and the alkaline earth metals are soluble.
7. The hydroxides of sodium, potassium, ammonium, and barium are soluble; calcium and strontium hydroxide are slightly soluble.

Vapor Pressure of Water at Various Temperatures

Temp. (°C)	Pressure (mm Hg)	Temp. (°C)	Pressure (mm Hg)
15	12.7	27	26.5
16	13.6	28	28.1
17	14.5	29	29.8
18	15.4	30	31.6
19	16.4	31	33.4
20	17.4	32	35.4
21	18.5	33	37.4
22	19.7	34	39.6
23	20.9	35	41.9
24	22.2		
25	23.5		
26	25.0		

Composition of Commercial Acids and a Base

Acid or base	Specific gravity	Percentage by mass	Molarity	Normality
Hydrochloric	1.19	38	12.4	12.4
Nitric	1.42	70	15.8	15.8
Sulfuric	1.84	95	17.8	35.6
Acetic	1.05	99	17.3	17.3
Aqueous ammonia	0.90	28	14.8	14.8

Generation of Common Gases

Oxygen. Oxygen is often prepared in the laboratory on a small scale by heating potassium chlorate, $KClO_3$, to about 50°C above its melting point. When manganese dioxide, MnO_2, is mixed with the chlorate, decomposition is rapid, even at 100°C below its melting point. The manganese dioxide can be reclaimed chemically unchanged after the reaction is completed; it has acted as a positive (accelerating) catalyst by having allowed decomposition of the chlorate to occur more easily. Preparing oxygen by heating potassium chlorate can be very dangerous. **Caution: Use goggles and/or an acrylic plastic shield. Explosions can occur when combustible material such as carbon, sulfur, or rubber comes in contact with fused (melted) potassium chlorate.** A safer method of producing oxygen involves the catalyzed decomposition of hydrogen peroxide.

● METHOD A. Heat strongly 2 g of manganese dioxide in a crucible for 2 or 3 minutes to make sure that it is free of combustible material. Allow it to cool and then transfer it to a Pyrex test tube. Add to the test tube 4 g of potassium chlorate and mix the contents thoroughly by shaking the closed test tube. Support the test tube as shown in Fig. 1 and wipe any loose powder from the inside of the test tube near its mouth. Fit the test tube with a one-hole stopper carrying a short glass tube. Connect this by rubber tubing to a long glass tube bent upward at the end and leading into a pneumatic trough of water. Fill four gas bottles with water, and invert them in the trough. No air bubbles should remain in the bottles. Insert the delivery tube into the neck of one of the bottles and begin heating the mixture contained in the test tube; hold the burner in your hand and do not overheat the mixture. Control the

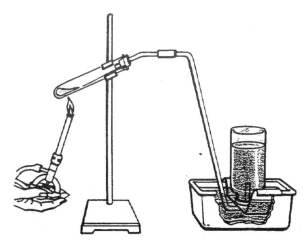

Figure 1 Assembly used in producing and collecting oxygen prepared by heating potassium chlorate.

heating so that the oxygen is evolved at a rate convenient for collection; if the evolution of gas becomes too rapid, as evidenced by the appearance of a white vapor, remove the flame for a moment. Never allow the test tube to cool while the delivery tube is under water; water will be forced back into the test tube as the volume of the gas contracts on cooling, and the cold water may cause the tube to crack.

When the first bottle is full of gas, fill it with water again; the first bottle of gas is discarded because it also contains the air in the system. Collect four bottles of oxygen and cover each with a glass plate before removal from the water trough. When the glass plate is in place, turn the bottle upright. The equation for the thermal decomposition of potassium chlorate is

$$2KClO_3 + \triangle \xrightarrow{MnO_2} 2KCl + 3O_2 \uparrow$$

⊙ METHOD B. Assemble a generator like that in Fig. 2, using either a 125-mL Erlenmeyer flask or a 120-mL wide-mouthed bottle. The thistle tube should be of polyethylene; if a glass thistle tube must be used, exercise caution in inserting it into the stopper (see Part 1B2). Clamp the generator on a ring stand. Place about 2 g of manganese dioxide and 25 mL of water in the generator and fit the stopper assembly tightly in place. Obtain about 60 mL of *freshly prepared* 6% hydrogen peroxide (not 3%). Fill four 250-mL wide-mouthed bottles with water and invert them in water contained in a pneumatic trough. Insert the delivery tube into one of the bottles, swirl the contents of the generator, and add 5 mL of the hydrogen peroxide through the thistle tube. Discard the first bottle of gas collected and then fill all four bottles with oxygen. Add the

hydrogen peroxide in 5-mL portions whenever the gas flow becomes too slow. Release the clamp and occasionally shake the generator gently. Slip a glass plate under the mouth of each bottle and turn the bottle upright. The first bottle of oxygen is discarded. The equation for the catalyzed decomposition of hydrogen peroxide is

$$2H_2O_2 \xrightarrow{MnO_2} 2H_2O + O_2 \uparrow + \triangle$$

Hydrogen. Hydrogen can be prepared conveniently in the laboratory by causing an active metal such as zinc to react with an acid such as sulfuric acid. The hydrogen of the sulfuric acid, H_2SO_4, is replaced by zinc, and the salt, zinc sulfate, $ZnSO_4$, is formed. Such chemical changes are known as displacement reactions. Set up the apparatus shown in Fig. 3. The thistle tube should be of polyethylene; if a glass thistle tube must be used, exercise care to avoid injury should the glass break (see Part 1B2). Add about 15 g of mossy zinc to the 150-mL widemouthed bottle and fit the stopper tightly in place. The end of the thistle tube should come within 3 mm of the bottom of the bottle. Connect the glass delivery tube by a 20-cm length of rubber tubing to a short piece of glass tubing that has been bent as shown in the figure.

Caution: Make sure there are no flames within 3 feet of the hydrogen generator. Mixtures of hydrogen and air are violently explosive. Wrap the generator tightly with a towel in order to prevent glass from flying in case of an explosion.

Add 100 mL of 3 M sulfuric acid and 3 drops of 0.01 M copper(II) sulfate solution (prepare from 0.1 M) through the thistle tube. The apparent catalytic effect of the copper(II) sulfate solution is due

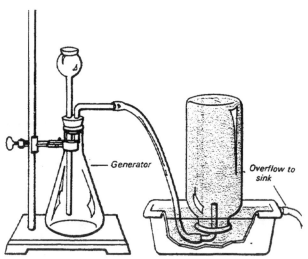

Figure 2 Apparatus used to generate and collect oxygen from hydrogen peroxide.

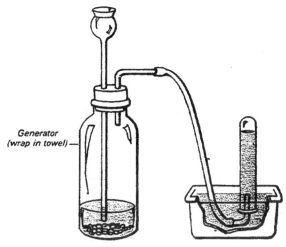

Figure 3 Apparatus used to generate and collect hydrogen.

to metallic copper that is displaced from solution by the more active zinc. Hydrogen bubbles form more easily on copper than on zinc surfaces.

When hydrogen gas is being evolved rapidly, collect a test tube full of it, place a finger over the mouth of the test tube, and carry the test tube with its mouth down to a flame. With the test tube still inverted, uncover the mouth and cautiously extend the test tube toward the flame.

Caution: Never have a flame close to a hydrogen generator.

If the gas in the test tube explodes when ignited by the flame, a mixture of hydrogen and air is indicated. Collect samples of hydrogen at 2-minute intervals until a sample is obtained that burns quietly. Collect four test tubes of hydrogen over water, close them with stoppers, and store them mouth downward.

The equation for the preparation of hydrogen by the reaction of zinc with sulfuric acid is

$$Zn + 2H^- + SO_4^{2-} \longrightarrow Zn^{2+} + SO_4^{2-} + H_2 \uparrow$$

Chlorine. Caution: Chlorine is a poisonous and irritating gas. All experiments in which chlorine is generated must be carried out in the hood. Place 20 g of powdered manganese dioxide in a 250-mL flask fitted with a two-hole stopper equipped with a thistle tube and delivery tube (Fig. 4). The thistle tube should be of polyethylene; if a glass thistle tube must be used, exercise care should the glass break (see Part 1B2). The thistle tube must reach nearly to the bottom of the flask. Make the rubber connection between the parts of the delivery tube as short as possible. Support the receiving bottle on wooden blocks (or on a Lab/Lift if available), so that a bottle can be removed and replaced with another without disturbing the ap-

paratus. Run the delivery tube into the bottle through a hole punched in a piece of cardboard. Ask the instructor to check the apparatus before proceeding.

Add 20 mL of concentrated hydrochloric acid through the thistle tube and warm the flask gently. Fill the desired number of bottles with the gas, judging by color (greenish-yellow) when each bottle is full. It may be necessary to add 20 mL more of acid during this time. Cover each bottle immediately with a glass plate; keep them in the hood. If chlorine water is also desired, finally bubble the gas through 50 mL of water in a beaker for a few minutes.

When enough product has been obtained, fill the generator with water by pouring the water through the thistle tube; wash the contents of the generator into the hood sink or other disposal places designated by the instructor.

The equation for this laboratory method for preparing chlorine is

$$MnO_2 + 2Cl^- + 4H^+ \longrightarrow Mn^{2+} + Cl_2 \uparrow + 2H_2O$$

Bromine. Caution: Bromine is very poisonous and irritating to skin and nasal passages. All experiments in which bromine is generated must be carried out in the hood. If bromine should come into contact with the skin, the affected part should be washed immediately with a saturated solution of sodium thiosulfate ("hypo") or with ethyl alcohol, then rinsed with copious quantities of water. Mix thoroughly 0.5 g of sodium bromide and 1 g of manganese dioxide. Place the mixture in a test tube and add about 3 mL of dilute sulfuric acid prepared by pouring concentrated sulfuric acid cautiously into an equal volume of water. Warm the tube gently if necessary; the properties of bromine vapor may be noted.

If the gas is to be collected and condensed, or bromine water prepared, a setup similar to that in Fig. 4 must be used. Observe the same precautions cited for the preparation of chlorine.

The net equation for the laboratory method for preparing bromine is

$$MnO_2 + 2Br^- + 4H^+ \longrightarrow Mn^{2+} + Br_2 + 2H_2O$$

List of Materials for Exercises 1–45

Estimated for 100 students; no allowance for waste. Omitted from the list are (1) equipment included in locker sets, (2) items commonly available in all laboratories, (3) unknowns, and (4) common acids and bases. Suitable unknowns for those exercises requiring them are given in the *Instructor's Guide*, Chapter 6.

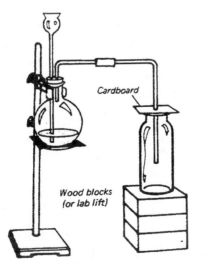

Cardboard

Wood blocks
(or lab lift)

Figure 4   Apparatus used to prepare chlorine in the laboratory.

Exercise 1. 1-q containers (or 32-oz bottles), objects of known mass expressed and labeled in a fraction of a pound, metal samples (in form of shot, granules, or pieces) or other suitable samples for density determinations.

Exercise 2. 300 mL ethanol, 10-mL graduated cylinders, 10-mL volumetric (or graduated) pipets, safety pipet fillers, unknowns (10-g solid unknown or 20-mL liquid unknown per student).

Exercise 3. 50 g copper(II) nitrate, 10 ft magnesium ribbon, 20 ft platinum or nichrome wire, 100 g copper metal sheet, 25 mL 0.1 M barium chloride, 25 mL 0.1 M sodium sulfate, 100 g calcium carbonate, 300 mL 0.1 M sodium iodide, 300 mL clorox bleach, 80 g iodine, 1 liter dichloromethane, 100 g powdered antimony.

Exercise 4. 500 g mixture for separation (see Instructor).

Exercise 5. 30 g magnesium ribbon, 60 g copper wire or turnings, 200 g powdered sulfur.

Exercise 6. 300 g various hydrated salts (identified to student only by *anhydrous* salt formula, not hydrate formula); e.g., barium chloride-2-hydrate, calcium chloride-2-hydrate, calcium chloride-6-hydrate, manganese sulfate-1-hydrate, magnesium sulfate-7-hydrate, zinc sulfate-7-hydrate.

Exercise 7. 150-mL aluminum tumblers (or flexible aluminum scrap), cardboard, 600-mL beaker and towels for insulation *or* foam plastic cups, 200×25-mm test tubes, 100 g metal (per student) as pellets or granules for specific heat determination, 50-mL volumetric pipets (preferable) *or* 50-mL graduated cylinders.

Exercise 8. 100 6-cm squares of aluminum foil, rubber bands, 1 liter 2-propanol, unknown (10 mL per student—optional).

Exercise 9. 200 200-mL foam plastic cups; sheet foam plastic; 10 liters each (based on 100 students) of 1.00 M sodium hydroxide and hydrochloric acid; 5 liters each (based on 100 students) of 1.00 M acetic acid and aqueous ammonia. (The solutions are to be standardized and adjusted to be 1.00 M within 1%.)

Exercise 10. Helium and hydrogen discharge tubes; spectroscope and accessories; 1 liter 0.1 M (or more concentrated) sodium chloride, potassium chloride, calcium chloride, strontium chloride, and barium chloride; platinum or nichrome wire for flame tests (or spray bottles), cobalt glass unknown (10 mL per student).

Exercise 11. 50 g sodium, 50 g calcium (or lithium), 40 ft magnesium ribbon, 200 copper strips (1×2 cm), 20 ft aluminum wire, 200 tin-foil strips (1×2 cm), 20 ft iron wire, 100 zinc strips (1×2 cm), sandpaper, 1 liter 0.1 M tin(II) chloride, 1 liter 0.1 M copper(II) sulfate, 1 liter 0.1 M silver nitrate, 20 mL mercury.

Exercise 12. Ball-and-stick molecular models.

Exercise 13. 500 mL 0.3% oleic acid in ethanol, 200 mL ethanol, lycopodium powder (or fine chalkdust, talcum powder, or finely divided sulfur); 15-cm watch glasses, small metric rulers, gummed tape, distilled water.

Exercise 14. Boyle's law apparatus.

Exercise 15. (per student or group) Soft-glass capillary tubing, 2-mm I.D. $\times$ 65 mm long (sealed on one end); 200-mL tall-form Pyrex beaker; heavy mineral oil; split cork; wire stirrer; small metric ruler; rubber bands.

Exercise 16. (per student or group) One or two Pyrex 10-mm I.D. $\times$ 60-cm-long glass tubes, cotton, 2 fitted corks/tube, stopwatch.

Exercise 17. 600 mL ethanol, 600 mL glycerol, 600 mL hexane, 1 kg sodium chloride, 200×25-mm test tubes, stiff wire for stirrers, 1 kg potassium nitrate, 25 10-mL graduated pipets, unknown (10 g per student).

Exercise 18. 1-liter 4.8 M acetamide (28.4%); 1 liter 4.8 M sodium chloride (28.1%); 20 conductivity apparatuses; 50 25-mL beakers; 1 liter each of the following 0.1 M solutions: copper(II) sulfate, sucrose, acetic acid, magnesium nitrate, sodium chloride; (hydrogen chloride in dry toluene); 1 liter ethanol; 1 liter glycerol; 200 g potassium nitrate; 200 g acetamide; 24 ammeters of suitable range; 500 mL of 0.01 M solutions of hydrochloric acid and sodium hydroxide; 500 mL of 0.10 M solutions of sulfuric acid and barium hydroxide; 500 mL of 0.10 M solutions of acetic acid and ammonia.

Exercise 19. 200×25-mm test tubes, stiff wire for stirrers, 3 kg diphenyl, 400 g naphthalene, 2 liters heptane, capillary tubes, unknowns (5 mL liquid unknown or 5 g solid unknown per student).

Exercise 20. 10 sealed tubes of nitrogen dioxide, 500 mL 0.1 M ammonium thiocyanate, 100 mL 0.1 M potassium thiocyanate, 200 mL 1 M ferric nitrate, 200 mL 1 M potassium thiocyanate, 100 mL 0.1 M silver nitrate, 20 mL 0.1 M disodium hydrogen phosphate.

Exercise 21. 50 50-mL eudiometers (gas-measuring tubes), 10 stopwatches, 270 g granulated zinc.

Exercise 22. (See Procedure, Exercise 22.) 150 mL potassium iodate solution and 50 mL starch-sulfite solution per student or group, 10-mL pipets (2 per student or group), stopwatches.

Exercise 23. 800 small test tubes; 400 medicine droppers; 2 liters each of 0.1 M solutions of potassium hyroxide, sodium sulfide, silver nitrate, cupric nitrate, ferric nitrate, barium chloride, sodium chloride, cupric chloride, sodium iodide, sodium sulfate, sodium carbonate, sulfuric acid, hydrochloric acid, nitric acid; 2 liters 0.5 M aqueous ammonia, unknowns (5 *coded* test tubes containing unknowns per student).

Exercise 24. 50 50-mL burets, 1 kg sodium hydroxide pellets, 10 liters standard 0.1 N hydrochloric acid, 100 mL phenolphthalein indicator (1%), 4 liters white vinegar, 25 10-mL pipets (preferably volumetric).

Exercise 25. pH meters, 10-mL volumetric pipets, 100-mL volumetric flasks, burets, magnetic stirrers, 0.1 M solutions of various unknown acids and their salts, 2 liters 0.1 M sodium chloride, 1 liter 1 M acetic acid, 1 liter 1 M sodium acetate, standardized 0.2 M sodium hydroxide, phenolphthalein, 1 liter 0.1 M hydrochloric acid, 1 liter 0.1 M sodium hydroxide, unknown acid solution (40 mL per student).

Exercise 26. 50 mL methyl orange indicator (1%); 50 mL phenolphthalein indicator (1%); 50 mL bromthymol blue indicator (1%); 1.5 liters 0.1 M acetic acid; 100 g sodium acetate; 200 mL 0.1 M aqueous ammonia; 100 g ammonium chloride; 500 mL each of 1 M solutions of sodium chloride, sodium acetate, ammonium chloride, ammonium acetate; 500 mL 0.1 M sodium bisulfite; 500 mL 0.1 M lead acetate; 500 mL 0.1 M sodium chloride; 1 liter 0.1 M hydrochloric acid; 500 mL 0.1 M sodium carbonate; 1 liter 0.1 M calcium nitrate; 1 liter 0.1 M sodium oxalate.

Exercise 27. 200 × 25-mm test tubes, 50 electrodes (stainless steel, platinum, or nichrome), 10 voltmeters (0–6 V dc), 5 liters 0.1 M iron(II) ammonium sulfate solution acidified with sulfuric acid, 1 liter 0.1 M potassium permanganate, 4 liters saturated sodium sulfate, cotton.

Exercise 28. 50 burets, 400 g potassium permanganate, 100 1-liter amber narrow-mouthed bottles, 10 liters standard 0.1 N sodium oxalate, 50 50-mL volumetric pipets, safety pipet filters, 20 250-mL volumetric flasks, 600 mL 85% phosphoric acid, unknowns (3 g per student). (See *Instructor's Guide* if standard oxalate solution is not available.)

Exercise 29. 50 copper strips (1 × 3 cm), 50 zinc strips (1 × 3 cm), 4 liters each of 1 M solutions of copper(II) sulfate and zinc sulfate, 50 porous porcelain cups (50-mL volume), 20 voltmeters (0–6 V dc), 100 mL 0.1 M sodium iodide, 100 g soluble starch, 20 mL phenolphthalein indicator (1%), dry cell or storage battery, 50 carbon electrodes, 3.5 liters special nickel solution for electroplating (see Procedure), copper wire.

Exercise 30. Samples of various kinds of natural water, burets, 100 mL 0.1 N sodium thiosulfate, 10 liters standardized 0.02 N sodium hydroxide (carbonate-free), methyl orange indicator (1%), phenolphthalein indicator (1%), pH meters.

Exercise 31. Samples of hard water, 100-mL buffer (6.75 g ammonium chloride dissolved in 57.0 mL concentrated ammonia and diluted to 100 mL), indicator (0.5 g eriochrome black T mixed with 4.5 g hydroxylamine hydrochloride and dissolved in 100 mL ethanol), standardized EDTA (26.7 g of the sodium salt and 0.67 g $MgCl_2$ dissolved in 5 liters of water—standardized against $CaCl_2$).

Exercise 32. Samples of clear natural water, spectrophotometer, 50-mL volumetric pipets, necessary glassware washed with hot dilute hydrochloric acid, ammonium molybdate reagent [31.4 g $(NH_4)_6Mo_7O_{24} \cdot 4H_2O$ in 200 mL water, to which is cautiously added a cooled solution of 250 mL sulfuric acid in 400 mL water; finally, mix 3.4 mL concentrated nitric acid with this solution and dilute to 1 liter], aminonaptholsulfonic acid reagent (weigh separately 0.75 g 1-amino-2-naphthol-4-sulfonic acid, 42 g anhydrous sodium sulfite, and 70 g anhydrous sodium metabisulfite, $Na_2S_2O_5$. Grind the sulfonic acid with a small portion of the $Na_2S_2O_5$ powder. Dissolve the remaining salts in about 900 mL of water. Dissolve the finely ground sulfonic acid in this mixture and dilute to 1 liter.

Store in brown glass-stoppered bottle in cool place; the reagent will keep for over 4 months), stock phosphate solution (0.7165 g anhydrous potassium dihydrogen phosphate, KH_2PO_4, per liter), standard phosphate solution (made by diluting 100 mL of stock solution to 1 liter; it contains 50 μg phosphate per milliliter).

Exercise 33. Provided sources of O_2 and/or H_2; wood splints; 50 g red phosphorus; steel wool; 20 ft magnesium ribbon; 50 g mercury(II) oxide, lead oxide, ferric oxide, silicon dioxide, and calcium oxide; 60 g manganese oxide; 100 g potassium chlorate; 10 g cupric oxide.

Exercise 34. 4 liters ascorbic acid stock solution (500 mg/L; must be made up fresh the day of the lab), 4 liters ascorbic acid standard solution (0.20 mg/mL; made by diluting 200 mL of stock to 500 mL), 1 liter each of dilutions of stock solutions for unknowns (0.10 mg/mL; 0.20 mg/mL; 0.30 mg/mL; 0.40 mg/mL), 2 liters glacial acetic acid, 10 liters 2,6-dichloro-indophenol dye solution (350 mg/liter; solution must be filtered), burets, buret brushes, 10-mL pipets, steam baths, fruit juice (to be supplied by student).

Exercise 35. 100 mL each of octane and cyclohexene, 100 mL 0.1 M potassium permanganate, 100 mL bromine water, 100 mL each of ethanol, *n*-butyl alcohol, benzyl alcohol, *sec*-butyl alcohol, *tert*-butyl alcohol, and isopentyl alcohol, 15 g sodium, 100 g benzoic acid, 100 g acetic acid, 100 g naphthylamine, 300 mL 0.1 M silver nitrate, 200 mL 0.1 M sodium bromide, 200 mL *n*-butyl bromide, 30 ft copper wire, 50 mL dichloromethane, 2 liters Fehling's solutions A and B, 50 mL 5% formaldehyde solution, 200 mL acetone, 100 mL each of 5% solutions of sucrose, glucose, maltose, lactose, and fructose.

Exercise 36. Ball-and-stick models or plastic Minit model kits (framework and/or space-filling model kits if available), small metric rulers.

Exercise 37. 5 vials each of cinnamic aldehyde, benzaldehyde, methyl benzoate, methyl butyrate, and vanillin; 20 10-mL graduated pipets, 15 g salicylic acid, 30 mL methanol, 500 mL ethanol, 25 g beta-naphthol, 1 kg tallow, cottonseed oil or lard, 10 liters saturated sodium chloride solution (100 g castile soap), 100 g synthetic detergent, 300 g sodium hydroxide, 50 mL cottonseed oil or kerosene, 100 mL 0.1 M calcium nitrate, 100 mL 0.1 M magnesium sulfate, 25 mL phenolphthalein indicator (1%), (100 g commercial soap such as Dial or Ivory—optional).

Exercise 38. 100 50-mL Erlenmeyer flasks, 200 g salicylic acid, 500 mL acetic anhydride, melting-point capillaries, 2 liters methanol, oil for oil baths (silicone oil; e.g., Dow Corning 550 is suggested), commercial aspirin, 100 mL 1% ferric chloride.

Exercise 39. Unknowns (50 g per student: fructose, glucose, lactose, maltose, starch, or sucrose), 1.5 liters Fehling's solution A (69.28 g powdered reagent-grade cupric sulfate pentahydrate dissolved in water and diluted to 1 liter), 1.5 liters Fehling's solution B (692 g sodium potassium tartrate and 200 g sodium hydroxide dissolved in water; dilute at room temperature to 2 liters), 150 mL 1% fructose, 1.2 liters 1% glucose, 100 mL 1% sucrose, 500 mL 1% starch, 500 mL 1% lactose, 200 mL dilute iodine solution, 1.5 liters Barfoed's reagent (dissolve 133 g powdered cupric acetate in 2 liters water; filter if necessary and to each 2 liters of filtrate add 50 mL 38% acetic acid. For the latter, dilute 18 mL glacial acetic acid with water to 50 mL), 3.6 liters Seliwanoff's reagent [add 3.5 mL 0.5% fresh resorcinol (1, 3-dihydroxy benzene) solution to 12 mL concentrated hydrochloric acid and dilute to 35 mL with distilled water], 45 g glucose, 45 g lactose.

Exercise 40. 100 200-mL distilling flasks and distillation setups, heating mantles (if available), 50 g fat, 2.5 liters 0.50 N potassium hydroxide in 95% ethanol, 2.5 liters 95% ethanol, 2.5 liters 0.50 N hydrochloric acid, phenolphthalein indicator (1%).

Exercise 41. Green or spinach leaves, sand, 1 liter of a solution of 400 mL of methanol and 850 mL of ligroin (bp 40–60°C), 20 m of 1.3-cm chromatographic paper strip, melting-point capillaries, 100 200 × 25-mm test tubes, 500-mL wide-mouth Erlenmeyer flasks, a commercial food color set (red, blue, green, and yellow samples), Scotch tape, 1.5 liters of 2-propanol, plastic wrap, silica gel TLC film (cut in 5 × 10-cm sections), 1 liter of developing solvent (3 volumes of toluene and 1 volume each of ether, acetone, and ethanol), 50 mL each of pheno, catechol, resorcinol, and pyrogallol, and a mixture of equal amounts of these (all 6% w/v in ethanol); crystalline iodine.

Exercise 42. 1-liter beakers, 2 liters of 2% aqueous ammonia, 4 liters of 2-propanol, aluminum or plastic foil, melting-point capillaries, 100 mL each of solutions of leucine, aspartic acid, tyrosine, glycine (all 0.05 M in 0.05 M hydrochloric acid), 100 mL 2% ninhydrin in ethanol, 100 12 × 22-cm sheets of chromatographic paper, 100 circles of chromatographic paper, 100 Petri dishes, developing chambers (see Procedure, Exercise 42, Part 3), 1 liter of developing solution (shake together 400 mL 1-butanol, 100 mL acetic acid, and 500 mL water; use only the upper layer), 200 cellulose-coated TLC strips (without indicator), 1 liter Visualizer Solution I (800 mL 0.2% ninhydrin in ethanol, 40 mL 2, 4, 6-trimethylpyridine, 160 mL acetic acid), 300 mL Visualizer Solution II (1% solution of cuprous nitrate trihydrate in anhydrous ethanol), coded unknowns (about 10 mL per student—optional).

Exercise 43. 10-mL volumetric pipets, safety pipet fillers, 25-mL volumetric flasks, spectrophotometers and accessories, 6 liters standardized 5×10^{-2} M ferric nitrate, 6 liters standardized 5×10^{-4} M potassium thiocyanate.

Exercise 44. 200 g aluminum scrap, 6 liters 1.5 M potassium hydroxide, glass wool, 2 liters 9 M sulfuric acid, 3 liters ethanol.

Exercise 45. 1 kg sodium bicarbonate, 1 kg USP pure baking soda.

List of Reagents for Qualitative Analysis

GROUP I *Reagents*

Hydrochloric acid, HCl, 6 M
Nitric acid, HNO_3, 4 M
Aqueous ammonia, $NH_3 + H_2O$, 4 M
*Potassium chromate, K_2CrO_4, 1 M

GROUP II *Reagents (in addition to those listed for Group I)*

Aqueous ammonia, $NH_3 + H_2O$, 6 M
Hydrochloric acid, HCl, 1.0 M
Thioacetamide, CH_3CSNH_2, 5% solution
Sodium hydroxide, NaOH, 4 M
Ammonium nitrate, NH_4NO_3, 1 M
Sulfuric acid, H_2SO_4, 4 M
Acetic acid, CH_3CO_2H, 1 M
Oxalic acid, $H_2C_2O_4$, 1 M
Ammonium acetate, $NH_4CH_3CO_2$, 1 M
Aqueous ammonia, $NH_3 + H_2O$, 15 M
*Potassium hexacyanoferrate(II), $K_4Fe(CN)_6$, 0.1 M
Potassium cyanide, KCN, 1 M
Hydrogen peroxide, H_2O_2, 3% solution
*Magnesia mixture: Dissolve 50 g of $MgCl_2 \cdot 6H_2O$ and 70 g of NH_4Cl in 400 mL of water. Add 100 mL of 15 M aqueous ammonia and dilute to 1 liter. Filter.
*Aluminum wire, Al
Mercury(II) chloride, $HgCl_2$, 0.2 M
Sodium hypochlorite, NaOCl, 5% solution
$SnCl_2$, 0.4 M
$SnCl_2$, 1 M

GROUP III *Reagents (in addition to those listed for Groups I and II)*

Hydrochloric acid, HCl, 12 M
Hydrochloric acid, HCl, 1 M
Nitric acid, HNO_3, 14 M
*Dimethylglyoxime, 1% solution: Dissolve 10 g in 1 liter of alcohol
Acetic acid, CH_3CO_2H, 4 M
*Ammonium thiocyanate, NH_4SCN, solid
Acetone, $(CH_3)_2CO$
*Sodium fluoride, NaF, 1 M
*Potassium nitrite, KNO_2, solid
*Sodium nitrite, $NaNO_2$, 1 M
*Sodium bismuthate, $NaBiO_3$, solid
*Aluminum reagent: Dissolve 1 g of the ammonium salt of aurin-tricarboxylic acid in 1 liter of water
Lead acetate, $Pb(CH_3CO_2)_2$, 0.1 M
Ammonium carbonate, $(NH_4)_2CO_3$, 1 M

GROUP IV *Reagents (in addition to those listed for Groups I–III)*

Ethanol, C_2H_5OH
*Ammonium oxalate, $(NH_4)_2C_2O_4$, 0.4 M

GROUP V *Reagents (in addition to those listed for Groups I–IV)*

*Ammonium sulfate, $(NH_4)_2SO_4$, 1 M
*Sodium reagent: Mix 30 g of $UO_2(CH_3Ca_2)_2 \cdot 2H_2O$ with 80 g of $Zn(CH_3CO_2)_2 \cdot 2H_2O$ and 10 mL of glacial acetic acid. Dilute the solution to 250 mL, let it stand for several hours, and filter. Use the clear solution.
*Sodium hexanitrocobaltate(III), $Na_3Co(NO_2)_6$. Dissolve 30 g of $NaNO_2$ in 97 mL of water; add 3 mL of glacial CH_3CO_2H and 3.3 g of $Co(NO_3)_2 \cdot 6H_2O$. Filter and use the clear solution. Add 1 to 5 drops CH_3OH per liter to stabilize.
*Disodium hydrogen phosphate, Na_2HPO_4, 1 M
*Magnesium reagent, *p*-nitro-benzene-azo-alpha-naphthol: Dissolve 0.25 g of this reagent and 2.5 g of NaOH in sufficient water to make 250 mL of solution.

*Fill reagent bottles only about one-fourth full of starred reagents.

Reagents for Anion Analysis
(in addition to those listed for the cation analysis)

---

*Sulfuric acid, H_2SO_4, 1.5 M
 Sodium carbonate, Na_2CO_3, 1.5 M
*Manganese(II) chloride, $MnCl_2$: Saturate 12 M HCl with $MnCl_2$.
*Potassium hexacyanoferrate(III), $K_3Fe(CN)_6$, a freshly prepared saturated solution
*Iron(III) chloride, $FeCl_3$, 0.1 M
 Calcium chloride, $CaCl_2$, 0.1 M
 Barium chloride, $BaCl_2$, 0.1 M
*Cobalt(II) nitrate, $Co(NO_3)_2$, 1 M
*Silver nitrate, $AgNO_3$, 0.1 M
*Bromine water, $Br_2 + H_2O$, saturated solution
*Barium hydroxide, $Ba(OH)_2$, saturated solution
*Iron(II) sulfate, $FeSO_4$, 0.1 M
*Sulfuric acid, H_2SO_4, 18 M
*Silver sulfate, Ag_2SO_4, solid (nitrate free)
*Potassium permanganate, $KMnO_4$, 0.002 M
*Silica, SiO_2, powdered
*Methanol, CH_3OH
*Iron(III) nitrate, $Fe(NO_3)_3$, 0.1 M
*Dichloromethane, CH_2Cl_2
*Potassium permanganate, $KMnO_4$, 0.1 M
*Lead acetate paper, filter paper moist with 0.1 M $Pb(CH_3CO_2)_2$

---

*Fill reagent bottles only about one-fourth full of starred reagents.

Preparation of Solutions of Cations

Stock Solutions. It is recommended that all stock solutions contain the cations in question at a concentration of 50 mg per milliliter. These stock solutions may be prepared by grinding to a powder the weight of salt given in the table on page 442 and adding enough water (or acid if specified) to make the volume 100 mL. Solution of the salts may be hastened by heating.

Known and Unknown Solutions. To prepare known or unknown solutions of cations, mix 20 mL of the stock solutions (40 mL of $AsCl_3$) of the cations desired and dilute the solution to 100 mL with water. This solution will contain 10 mg of cations per milliliter. This procedure allows for a maximum of five cations at a concentration of 10 mg of cations per milliliter. Dilution to 200 mL will allow a maximum of ten cations at a concentration of 5 mg per milliliter. Give each student about 1 mL of solution.

DISPOSAL OF WASTES*[1]

Heavy Metal Compounds

In general, these are precipitated and disposed of in a managed hazardous waste facility. The filtrates are washed down the drain.

1. Barium Compounds A solution of a barium salt is precipitated with 10% sodium sulfate. Although barium compounds are toxic, barium sulfate (a fine precipitate) is not considered a hazardous waste. The only occurrence in this manual of a barium salt is in Exercise 3, and barium sulfate is a product.

2. Lead, Cobalt, and Silver Compounds Precipitate as sulfides with 10% sodium sulfide. No lead or cobalt salts are used in the manual. If a silver salt solution is to be dicarded, use the method given here.

Solutions of Other Metals Used in This Manual

Prepare a *dilute* solution and pour it slowly down the drain with running water.

Mercury

Collect used or spilled mercury and save in a *closed* container for future purification. It is important to clean up *all* spilled mercury. Commercial mercury

---

*Consult the regional office of the U. S. Environmental Protection Agency in your area for specific, more complete details.
[1]Based on *Chemical Activities*, Teacher Edition. Copyright © 1988 by The American Chemical Society, pages 318–319.

Stock Solutions of Cations (50 mg of cations per milliliter)

Group	Ion	Formula of salt	Grams per 100 mL of solution
I	Ag^+	$AgNO_3$	8.0
	Pb^{2+}	$Pb(NO_3)_2$	8.0
	Hg_2^{2+}	$Hg_2(NO_3)_2$	7.0 (dissolve in 0.6 M HNO_3)
II	Pb^{2+}	$Pb(NO_3)_2$	8.0
	Bi^{3+}	$Bi(NO_3)_3 \cdot 5H_2O$	11.5 (dissolve in 3 M HNO_3)
	Cu^{2+}	$Cu(NO_3)_2 \cdot 3H_2O$	19.0
	Cd^{2+}	$Cd(NO_3)_2 \cdot 4H_2O$	13.8
	Hg^{2+}	$HgCl_2$	6.8
	As^{3+}	As_4O_6	3.3 (heat in 50 mL of 12 M HCl, then add 50 mL of water)
	Sb^{3+}	$SbCl_3$	9.5 (dissolve in 6 M HCl and dilute with 2 M HCl)
	Sn^{2+}	$SnCl_2 \cdot 2H_2O$	9.5 (dissolve in 50 mL of 12 M HCl; dilute to 100 mL with water; add a piece of tin metal)
	Sn^{4+}	$SnCl_4 \cdot 3H_2O$	13.3 (dissolve in 6 M HCl)
III	Co^{2+}	$Co(NO_3)_2 \cdot 6H_2O$	24.7
	Ni^{2+}	$Ni(NO_3)_2 \cdot 6H_2O$	24.8
	Mn^{2+}	$Mn(NO_3)_2 \cdot 6H_2O$	26.2
	Fe^{3+}	$Fe(NO_3)_3 \cdot 9H_2O$	36.2
	Al^{3+}	$Al(NO_3)_3 \cdot 9H_2O$	69.5
	Cr^{3+}	$Cr(NO_3)_3$	23.0
	Zn^{2+}	$Zn(NO_3)_2$	14.5
IV	Ba^{2+}	$BaCl_2 \cdot 2H_2O$	8.9
	Sr^{2+}	$Sr(NO_3)_2$	12.0
	Ca^{2+}	$Ca(NO_3)_2 \cdot 4H_2O$	29.5
V	Mg^{2+}	$Mg(NO_3)_2 \cdot 6H_2O$	52.8
	NH_4^+	NH_4NO_3	22.2
	Na^+	$NaNO_3$	18.5
	K^+	KNO_3	13.0

collection kits are available. To dispose of mercury in a waste facility, first convert it to a mercury sulfide by dusting with powdered sulfur.

Metals (other than mercury, sodium, etc.)

Wash and store for future use. Do not discard finely divided (powdered) metal in the solid waste container, which may contain paper.

Nonmetals

Do not discard iodine in the solid waste container.

Acids and Bases

Dilute small amounts by adding acids or bases to a larger amount of water and pour the mixture down the drain. Or neutralize acids with dilute aqueous ammonia or baking soda solution; neutralize bases with dilute hydrochloric acid or vinegar solution, and flush down the drain.

For spilled acids or bases, confine the area of the spill, neutralize, and mop up.

Organic Compounds

1. **Alcohols and Organic Acids.** Dilute with water and wash slowly down the drain.

2. **Hydrocarbon Solvents.** Do not discard in the sink. Consult your school maintenance person for the correct procedure.

DESCRIPTION OF HAZARDOUS CHEMICALS

The chemicals used in this manual were screened for safety. They are safe to use with reasonable care, in the amounts specified, and with attention to necessary caution warnings, which are emphasized in bold black type throughout the manual. The publication "School Science Laboratories, A Guide to Some Hazardous Substances," available from the U.S. Consumer Products Safety Commission,* classifies chemicals as explosives, flammables, carcinogens (known or possible human), carcinogens (animal), mutagens (possible gene-altering), toxics, corrosives, or irritants.

No *explosive* chemicals as such are called for in this manual. Of course, explosions can occur for a number of other reasons, usually because of negligence in operations.

Flammables such as organic liquids or hydrogen, when used, must never be in proximity to an open flame.

Known human *carcinogens* have already been eliminated in earlier editions; e. g., benzene and carbon tetrachloride (toluene and dichloromethane were substituted).

Animal *carcinogens* or *mutagens* must be handled with care. For example, caution is tressed in handling solid acetamide (Exercise 18), and minimal use of formaldehyde (Exercise 35) is directed.

Many chemicals are *toxics* (whether inhaled, absorbed, or ingested). Mercury (and its vapor) is one of high toxicity (see Disposal of Wastes). Unused mercury must be kept stored in a closed container.

Corrosives, such as higher concentrations of sulfuric acid, react with skin and result in chemical burns (see Part 1B, Safety in the Laboratory).

Many chemicals are *irritants,* and induce local inflammatory reactions. Avoid inhaling dust and fumes, and exposures to the skin (see Part 1B, Safety in the Laboratory).

---

*U. S. Consumer Products Safety Commission, Office of the Secretary, Washington, D. C. 20207.

CATION UNKNOWN REPORT

No.	Substance	Reagent	Result	Inference or conclusion	Precipitate or residue	Centrifugate or solution
1						
2						
3						
4						
5						
6						
7						
8						
9						
10						
11						
12						
13						
14						

NAME _____ SECTION _____ DATE _____

INSTRUCTOR'S APPROVAL _____ IONS FOUND _____ GRADE _____

CATION UNKNOWN REPORT

No.	Substance	Reagent	Resi...	Inference or conclusion	Precipitate or residue	Centrifugate or solution
1						
2						
3						
4						
5						
6						
7						
8						
9						
10						
11						
12						
13						
14						

NAME _____ SECTION _____ DATE _____

INSTRUCTOR'S APPROVAL _____ IONS FOUND _____ GRADE _____

CATION UNKNOWN REPORT

No.	Substance	Reagent	Result	Inference or conclusion	Precipitate or residue	Centrifugate or solution
1						
2						
3						
4						
5						
6						
7						
8						
9						
10						
11						
12						
13						
14						

NAME _____ SECTION _____ DATE _____

INSTRUCTOR'S APPROVAL _____ IONS FOUND _____ GRADE _____

CATION UNKNOWN REPORT

No.	Substance	Reagent	Result	Inference or conclusion	Precipitate or residue	Centrifugate or solution
1						
2						
3						
4						
5						
6						
7						
8						
9						
10						
11						
12						
13						
14						

CATION UNKNOWN REPORT

No.	Substance	Reagent	Result	Inference or conclusion	Precipitate or residue	Centrifugate or solution
1						
2						
3						
4						
5						
6						
7						
8						
9						
10						
11						
'12						
13						
14						

CATION UNKNOWN REPORT

No.	Substance	Reagent	Result	Inference or conclusion	Precipitate or residue	Centrifugate or solution
1						
2						
3						
4						
5						
6						
7						
8						
9						
10						
11						
12						
13						
14						

NAME _____ SECTION _____ DATE _____

INSTRUCTOR'S APPROVAL _____ IONS FOUND _____ GRADE _____

CATION UNKNOWN REPORT

No.	Substance	Reagent	Result	Inference or conclusion	Precipitate or residue	Centrifugate or solution
1						
2						
3						
4						
5						
6						
7						
8						
9						
10						
11						
12						
13						
14						

NAME _____ SECTION _____ DATE _____

INSTRUCTOR'S APPROVAL _____ IONS FOUND _____ GRADE _____

ANION ELIMINATION CHART

Anions	Test 1	Test 2	Test 3	Test 4	Test 5	Test 6	Test 7	Test 8	Make confirmatory tests for
CO_3^{2-}									
SO_3^{2-}									
S^{2-}									
NO_2^-									
NO_3^-									
I^-									
F^-									
$C_2O_4^{2-}$									
SO_4^{2-}									
PO_4^{3-}									
BO_2^-									
Cl^-									
Br^-									

NAME _____ SECTION _____ DATE _____

INSTRUCTOR'S APPROVAL _____ IONS FOUND _____ GRADE _____

ANION ELIMINATION CHART

Anions	Test 1	Test 2	Test 3	Test 4	Test 5	Test 6	Test 7	Test 8	Make confirmatory tests for
CO_3^{2-}									
SO_3^{2-}									
S^{2-}									
NO_2^{-}									
NO_3^{-}									
I^{-}									
F^{-}									
$C_2O_4^{2-}$									
SO_4^{2-}									
PO_4^{3-}									
BO_2^{-}									
Cl^{-}									
Br^{-}									

NAME _____ SECTION _____ DATE _____

INSTRUCTOR'S APPROVAL _____ IONS FOUND _____ GRADE _____

ANION ELIMINATION CHART

Anions	Test 1	Test 2	Test 3	Test 4	Test 5	Test 6	Test 7	Test 8	Make confirmatory tests for
CO_3^{2-}									
SO_3^{2-}									
S^{2-}									
NO_2^-									
NO_3^-									
I^-									
F^-									
$C_2O_4^{2-}$									
SO_4^{2-}									
PO_4^{3-}									
BO_2^-									
Cl^-									
Br^-									

NAME _____ SECTION _____ DATE _____

INSTRUCTOR'S APPROVAL _____ IONS FOUND _____ GRADE _____

ANION ELIMINATION CHART

Anions	Test 1	Test 2	Test 3	Test 4	Test 5	Test 6	Test 7	Test 8	Make confirmatory tests for
CO_3^{2-}									
SO_3^{2-}									
S^{2-}									
NO_2^-									
NO_3^-									
I^-									
F^-									
$C_2O_4^{2-}$									
SO_4^{2-}									
PO_4^{3-}									
BO_2^-									
Cl^-									
Br^-									

NAME _____ SECTION _____ DATE _____

INSTRUCTOR'S APPROVAL _____ IONS FOUND _____ GRADE _____

ANION ELIMINATION CHART

Anions	Test 1	Test 2	Test 3	Test 4	Test 5	Test 6	Test 7	Test 8	Make confirmatory tests for
CO_3^{2-}									
SO_3^{2-}									
S^{2-}									
NO_2^-									
NO_3^-									
I^-									
F^-									
$C_2O_4^{2-}$									
SO_4^{2-}									
PO_4^{3-}									
BO_2^-									
Cl^-									
Br^-									

The Periodic Table of Elements (*long form*)

Atomic masses are rounded to three decimals, where available; atomic numbers appear above the symbols.

Metals / Transition metals / Nonmetals / Inner transition metals

Period	1 IA	2 IIA	3 IIIB	4 IVB	5 VB	6 VIB	7 VIIB	8 VIIIB	9 VIIIB	10 VIIIB	11 IB	12 IIB	13 IIIA	14 IVA	15 VA	16 VIA	17 VIIA	18 VIIIA
1	1 H 1.008																	2 He 4.003
2	3 Li 6.941	4 Be 9.012											5 B 10.811	6 C 12.011	7 N 14.007	8 O 15.999	9 F 18.998	10 Ne 20.180
3	11 Na 22.990	12 Mg 24.305											13 Al 26.982	14 Si 28.086	15 P 30.974	16 S 32.066	17 Cl 35.453	18 Ar 39.948
4	19 K 39.098	20 Ca 40.078	21 Sc 44.956	22 Ti 47.88	23 V 50.942	24 Cr 51.996	25 Mn 54.938	26 Fe 55.845	27 Co 58.933	28 Ni 58.693	29 Cu 63.546	30 Zn 65.39	31 Ga 69.723	32 Ge 72.61	33 As 74.922	34 Se 78.96	35 Br 79.904	36 Kr 83.80
5	37 Rb 85.468	38 Sr 87.62	39 Y 88.906	40 Zr 91.224	41 Nb 92.906	42 Mo 95.94	43 Tc (98)	44 Ru 101.07	45 Rh 102.906	46 Pd 106.42	47 Ag 107.868	48 Cd 112.411	49 In 114.818	50 Sn 118.710	51 Sb 121.75	52 Te 127.60	53 I 126.904	54 Xe 131.29
6	55 Cs 132.905	56 Ba 137.327	* (57–71)	72 Hf 178.49	73 Ta 180.948	74 W 183.85	75 Re 186.207	76 Os 190.23	77 Ir 192.217	78 Pt 195.08	79 Au 196.966	80 Hg 200.59	81 Tl 204.383	82 Pb 207.2	83 Bi 208.980	84 Po (209)	85 At (210)	86 Rn (222)
7	87 Fr (223)	88 Ra 226.025	† (89–103)	104 Rf (261)	105 Ha (262)	106 Sg (263)	107 Ns (262)	108 Hs (265)	109 Mt (266)	110 Uun (269)	111 Uuu (272)	112 Uub (277)						

*Lanthanide series

57 La 138.906	58 Ce 140.115	59 Pr 140.908	60 Nd 144.24	61 Pm (145)	62 Sm 150.36	63 Eu 151.965	64 Gd 157.25	65 Tb 158.925	66 Dy 162.50	67 Ho 164.930	68 Er 167.26	69 Tm 168.934	70 Yb 173.04	71 Lu 174.967

†Actinide series

89 Ac 227.028	90 Th 232.038	91 Pa 231.036	92 U 238.029	93 Np 237.048	94 Pu (244)	95 Am (243)	96 Cm (247)	97 Bk (247)	98 Cf (251)	99 Es (252)	100 Fm (257)	101 Md (258)	102 No (259)	103 Lr (260)